CROP PRODUCTION SCIENCE IN HORTICULTURE

Series Editors: Jeff Atherton, Senior Lecturer in Horticulture, University of Nottingham, and Alun Rees, Horticultural Consultant and Editor, *Journal of Horticultural Science*

This series examines economically important horticultural crops selected from the major production systems in temperate, sub-tropical and tropical climatic areas. Systems represented range from open field and plantation sites to protected plastic and glass houses, growing rooms and laboratories. Emphasis is placed on the scientific principles underlying crop production practices rather than on providing empirical recipes for uncritical acceptance. Scientific understanding provides the key to both reasoned choice of practice and the solution of future problems.

Students and staff at universities and colleges throughout the world involved in courses in horticulture, as well as in agriculture, plant science, food science and applied biology at degree, diploma or certificate level will welcome this series as a succinct and readable source of information. The books will also be invaluable to progressive growers, advisors and end-product users requiring an authoritative, but brief, scientific introduction to particular crops or systems. Keen gardeners wishing to understand the scientific basis of recommended practices will also find the series very useful.

The authors are all internationally renowned experts with extensive experience of their subjects. Each volume follows a common format covering all aspects of production, from background physiology and breeding, to propagation and planting, through husbandry and crop protection, to harvesting, handling and storage. Selective references are included to direct the reader to further information on specific topics.

Titles Available:
1. **Ornamental Bulbs, Corms and Tubers** A. R. Rees
2. **Citrus** F. S. Davies and L. G. Albrigo
3. **Onions and Other Vegetable Alliums** J. L. Brewster

Titles in Preparation:
Cucurbits R. W. Robinson
Tomatoes J. Atherton
Ornamental Bedding Plants A. M. Armitage
Carrots and Related Vegetable Umbelliferae V. E. Rubatzky, P. W. Simon and
C. F. Quiros
Coffee, Cocoa and Tea K. C. Willson
Bananas and Plantains J. C. Robinson
Horticultural Food Legumes G. D. Hill and J. Smartt
Temperate Zone Herbs L. E. Craker, J. E. Simon and B. Galambosi

CITRUS

Frederick S. Davies
Professor of Horticulture
Department of Horticultural Sciences
University of Florida
Gainesville, Florida, USA

and

L. Gene Albrigo
Professor of Horticulture
Citrus Research and Education Center
Lake Alfred, Florida, USA

CAB INTERNATIONAL

CAB INTERNATIONAL
Wallingford
Oxon OX10 8DE
UK

Tel: +44 (0)1491 832111
Fax: +44 (0)1491 833508
E-mail: cabi@cabi.org
Telex: 847964 (COMAGG G)

A catalogue record for this book is available from the British Library, London, UK.
A catalogue record for this book is available from the Library of Congress, Washington DC, USA.

ISBN 0 85198 867 9

First printed 1994
Reprinted 1998

Typeset by Solidus (Bristol) Limited
Printed in the UK at the University Press, Cambridge

CONTENTS

PREFACE

Citrus fruits have been cultivated and enjoyed for over 4000 years. Moreover, they are grown in nearly every country within 40° north–south latitude. This worldwide dissemination has been associated with many of the great explorations and conflicts in history including the conquests of Alexander the Great, the spread of Muhammedism and the explorations of Columbus, who brought citrus to the New World.

The intent of this book is to provide the reader with an overview of citriculture from a worldwide perspective. As a practical matter, we cannot discuss individual cultural programmes for every citrus-growing region, but instead we have tried to emphasize current theories and technological advances in citriculture, citing specific examples of how and where they are used. We have also included many current references and reviews on various aspects of citriculture for persons desiring more detail on a specific topic. However, due to space limitations we have by no means attempted to provide an exhaustive literature review.

The text begins with a discussion of major production areas and figures and current trends (Chapter 1). We then discuss the confusing and controversial taxonomic situation for *Citrus* and related genera emphasizing current biotechnological advances in the field. This is followed by a discussion of the major commercially important citrus species and cultivars and traditional and current techniques in citrus breeding (Chapter 2). In Chapter 3, the role of climatic factors in worldwide citrus production is emphasized. Climate has a pronounced effect on citrus yields, growth, economic returns and fruit quality. In Chapter 4 we cover the importance of rootstocks in citriculture and discuss the major rootstocks, their advantages and disadvantages. Plant husbandry including nursery practices, irrigation, fertilization, freeze protection, pruning and growth regulators is covered in Chapter 5. In Chapter 6 we review the major pest constraints, weeds, arthropods and diseases of citrus, emphasizing major problem areas and control measures. The final chapter covers postharvest quality, harvesting and handling of citrus fruits including importance, biotic and abiotic problems, as well as packinghouse and processing techniques. The information contained in the text should provide the reader with the most important basic information on citriculture, although certainly not all aspects have been addressed.

Several reviewers have made extremely helpful and constructive suggestions on improving the text including G.A. Moore, M.L. Roose, C. Lovatt, J.P. Syvertsen, C. McCoy, H. Browning, J. Burns, L. Timmer, R.F. Lee, W.S. Castle, G.E. Brown and H.K. Wutscher. Nevertheless, the text has undergone numerous revisions and we take full responsibility for any errors that may occur. We would also like to thank the series editors, J. Atherton and A. Rees, for their helpful suggestions and T. Hardwick for his editorial assistance and patience throughout the project. We also extend our sincere appreciation to Ms Tami Spurlin for the many hours she spent typing and editing the text.

Frederick S. Davies
L. Gene Albrigo

1

HISTORY, DISTRIBUTION AND USES OF CITRUS FRUIT

HISTORY

The general area of origin of *Citrus* is believed to be southeastern Asia, including that from eastern Arabia east to the Philippines and from the Himalayas south to Indonesia or Australia (Fig. 1.1). Within this large region, northeastern India and northern Burma were believed to be the centre of origin, but recent evidence suggests that Yunnan Province in south-central China may be as important due to the diversity of species found and the system of rivers that could have provided dispersal to the south (Gmitter and Hu, 1990). Extensive movement of the various types of citrus probably occurred within the general area of citrus origin from before recorded history. Many types of citrus are believed to have moved west to various Arabian areas, such as Oman, Persia, Media (Iran) and even Palestine, before Christ (Tolkowsky, 1938). Major types of edible citrus include citron, sour orange, lime, lemon, sweet orange, shaddock (pummelo), grapefruit, mandarin and kumquat. Cultivar characteristics, genetics and taxonomy of the major edible citrus species are presented in Chapter 2.

Citrons (*Citrus medica* L.) originated in the region from south China to India. The citron was found in Media when Alexander of Macedonia entered Asia (about 330 BC) and was subsequently introduced into the Mediterranean region. Other citruses were introduced to Italy during the early Roman Empire (27 BC–AD 284), but they are believed to have been destroyed at the end of this era. Controversy exists about whether citron is mentioned in the Bible, but it is clear that the Jewish religion was using it in their ceremonies by AD 50–150 (Webber, 1967).

Limes (*Citrus aurantifolia* Swingle) apparently originated in the east Indian archipelago. They were probably brought across the Sea of Oman by Arabian sailors and subsequently transported to Egypt and Europe.

Lemons (*Citrus limon* Burmann) are of unknown origin, possibly a hybrid between citron and lime creating an intermediate species (Chapot, 1975;

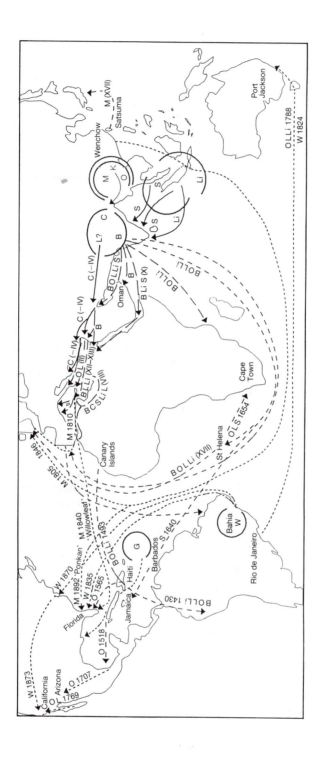

Fig. 1.1. The areas of origin of major citrus species and their paths of distribution (Chapot, 1975). Abbreviations: B, bigarade (*Citrus aurantium*); C, citron (*Citrus medica*); G, grapefruit (*Citrus paradisi*); K, kumquat (*Fortunella margarita*); L, lime (*Citrus aurantifolia*); Li, lemon (*Citrus limon*); M, mandarin (*Citrus reticulata*); S, shaddock (*Citrus grandis*); W, 'Washington' navel; BC, — —; AD 1–700, ————; AD 700–1492 (711: Arab occupation of Spain); – – , AD 1493–1700 (1493: second expedition of Christopher Columbus); ‧‧‧‧‧‧, after AD 1700 (first appearance of grapefruit). Centuries are given in roman numerals (minus sign indicates BC); years are given in arabic numerals.

Barrett and Rhodes, 1976). Citron is believed to be a more primitive species and limes and lemons are at least closely related (Barrett and Rhodes, 1976). Lemons are known to have been spread to North Africa and Spain about AD 1150 in connection with expansion of the Arabian Empire.

The area of origin of the sour orange (*Citrus aurantium* L.) is believed to be southeastern Asia, possibly India. Sour orange was introduced progressively westwards in the early centuries AD associated with Arab conquests until it reached North Africa and Spain in around AD 700.

Sweet oranges (*Citrus sinensis* [L.] Osbeck) originated in southern China and possibly as far south as Indonesia (Webber *et al.*, 1967). The sweet orange may have travelled a similar route as the citron and been introduced to Europe by the Romans. Greenhouses were developed during this time to protect the cold tender plants in pots in winter in the gardens of influential Roman families and were called orangeries, implying that oranges were a primary plant maintained by the Romans (Tolkowsky, 1938). Early introductions were apparently lost after the fall of the Roman Empire and oranges were reintroduced in around 1425 through the Genoese trade routes (Webber *et al.*, 1967). More selections were no doubt introduced during the extensive trading of the Venezia era, between the 15th and 17th centuries AD. The Portuguese brought superior selections of sweet oranges from China in about AD 1500. The great families of the Italian City States maintained large collections of lemons, oranges, etc. (paintings and descriptions of the cultivar collections of citrus and other fruits of the Medici family are on display in museums in Florence, Italy).

The 'Washington' navel orange originated in Bahia, Brazil, and is probably a mutation of 'Seleta' sweet orange. It was introduced to Australia (1824), Florida (1835) and California (1870) from Washington, DC, where it apparently received its name (Fig. 1.1). 'Washington' navel and many cultivars arising from mutations of it have been distributed worldwide (Davies, 1986a).

The shaddock or pummelo (*Citrus grandis* [L.] Osbeck) originated in the Malaysia and Indian archipelagos and is widely distributed in the Fiji Islands. Hybrids of shaddock were apparently found by crusaders in Palestine by AD 900 and were distributed to Europe and then to the Caribbean apparently by an East Indian ship's captain named Shaddock (Webber *et al.*, 1967).

Grapefruits appear to have originated as a mutation or hybrid of the shaddock in the West Indies, perhaps Barbados. The binomial *Citrus paradisi* Macf. was assigned to this species of questionable validity (see Chapter 2). Grapefruits were introduced into Florida, now the major producer worldwide, from the Caribbean in about 1809 by Don Phillippe probably from seed collected in Jamaica.

The area of origin for mandarins (*Citrus reticulata* Blanco) was probably Indo-China and south China with traders carrying selections to eastern India. Traditional production areas of this species have been in Asia. Mandarins were transported from Asia to Europe much later than other citrus; the 'Willowleaf' (*Citrus deliciosa* Tenole) was taken from China after 1805 to become the major species in the Mediterranean region and *C. reticulata* was introduced even later.

Movement of citrus to Africa from India probably occurred during AD 700–1400 and various citrus, particularly limes and oranges, were introduced to the Americas by the Mediterranean explorers and settlers centred in Hispaniola in the Caribbean (Spanish) and Bahia, Brazil (Portuguese). The movement of citrus through the Americas was aided by the development of missions by the Roman Catholic Church that established plantings of various fruit including citrus.

Minor use citruses such as edible kumquats (*Fortunella margarita* [Lour.] Swingle) from southern China and trifoliate orange (*Poncirus trifoliata* [L.] Raf.) from central and northern China are also important species, for freeze-hardy rootstocks. Other related species are being employed in new biotechnology-oriented breeding programmes (see Chapter 2).

DISTRIBUTION AND PRODUCTION

Major Production Areas

Currently, citrus is grown primarily between the latitudes 40°N to 40°S. More northern and southern locations of commercial production exist where temperatures are moderated by ocean winds. Major citrus-producing countries and their total production of various types of citrus as of 1989–1991 are shown in Table 1.1. The northern (Spain, Italy, Greece and Turkey) and southern (Morocco, Egypt, Israel, Tunisia, Lebanon and Algeria) Mediterranean regions, the northern (USA, Mexico, Belize) and southern (Brazil, Venezuela, Argentina and Uruguay) regions and associated islands (Cuba, Jamaica and Dominican Republic) of the American continents, China, Japan, South Africa and Australia are currently the major commercial production regions of citrus in the world. Brazil is now the largest producer of citrus worldwide having an industry oriented towards production of oranges for processing. The United States is the second largest producer of citrus and is the largest producer of grapefruit. Most production is geared toward processing in Florida and to fresh fruit production in California, Arizona and Texas. California is a major shipper of lemons and navel oranges, 61% of its orange plantings being navel oranges. Severe freezes during the 1980s caused considerable crop and tree losses in the United States. Recently, China, a major area of origin, has significantly increased production to become the third largest producer of oranges worldwide and the fourth in overall production (Table 1.1). This represents nearly a tenfold increase from 1969 and shifts emphasis to oranges from mandarin-type fruits which used to

Table 1.1. World citrus production (two-year average of 1989–1990 and 1990–1991 seasons). All values are expressed in 1000 × tonnes. (In some cases values for minor unlisted citrus-producing areas may have been included in the totals.)

Location	Total	Oranges	Mandarins	Lemons/limes	Grapefruits
World	67,967.9	48,568.4	8,580.9	6,404.7	4,413.9
Northern hemisphere	47,155.5	30,835.1	7,510.1	4,916.5	3,893.9
United States	9,987.6	7,163.0	262.7	638.1	1,923.9
Mediterranean region	17,297.7	10,807.6	3,301.8	2,476.7	711.6
Greece	1,186.8	875.7	74.9	229.3	–
Italy	3,233.3	2,025.0	509.0	691.0	8.3
Spain	4,862.8	2,613.4	1,519.9	708.0	21.5
Israel	1,279.5	720.7	125.8	39.0	394.1
Algeria	298.6	185.0	108.0	3.1	2.6
Morocco	1,251.8	944.8	285.1	15.0	7.0
Tunisia	276.7	172.6	36.6	15.6	52.0
Cyprus	313.0	144.6	–	55.5	103.0
Egypt	1,887.7	1,498.7	175.0	212.0	–
Lebanon	447.8	279.1	–	–	–
Turkey	1,492.5	804.1	328.5	320.5	39.5
Former USSR	178.0	178.0	–	–	–
Japan	2,423.0	239.5	2,181.5	–	–
Cuba	1,020.0	520.0	30.0	70.0	400.0
Mexico	3,303.5	2,300.0	183.5	711.0	109.0
China	5,511.4	4,711.9	385.0	149.5	265.0
Southern hemisphere	20,128.8	17,733.3	1,070.8	1,488.2	520.1
Argentina	1,585.0	750.0	245.0	405.0	185.0
Brazil	14,300.0	13,050.0	581.0	644.0	25.0
Uruguay	193.8	83.5	49.5	53.0	–
Venezuela	427.0	427.0	–	–	–
Australia	629.5	519.5	–	35.0	31.0
South Africa	874.0	691.0	–	60.0	123.0
Chile	–	–	–	75.0	–

Source: FAO (1991).

predominate (Spurling, 1969; Wen-cai, 1981). Many of the popular cultivars in China are local selections and are not grown in the western world. These include several cultivars of satsuma mandarin and over 30 orange cultivars. Spain is the fourth largest producer followed by Mexico, Italy, Japan, Egypt, Argentina, Turkey, Israel and Morocco. The relative ranking of these countries has varied over the last ten years, with production increasing in Mexico, Egypt and Argentina and decreasing in Japan. Production in Cuba also increased significantly from 1987 to 1991.

Table 1.2. Fresh citrus exports by principal countries (average of 1989–1990 and 1990–1991 seasons). All values are expressed in 1000 × tonnes. (In some cases values for minor unlisted citrus-growing areas may have been included in the totals.)

Location	Total	Oranges	Mandarins	Lemons/limes	Grapefruits
World	7,858.4	4,449.2	1,375.5	1,002.4	1,031.4
Northern hemisphere	6,951.3	3,790.3	1,360.0	890.2	910.9
United States	916.0	381.5	13.0	131.0	390.5
Mediterranean region	4,948.4	2,841.9	1,211.1	662.7	232.8
Greece	356.0	296.0	–	49.0	–
Italy	224.5	153.0	11.0	59.5	–
Spain	2,379.7	1,151.6	862.5	358.5	7.1
Israel	386.6	236.3	29.3	–	112.4
Algeria	3.0	2.0	1.0	–	–
Morocco	561.0	427.5	132.4	–	–
Tunisia	27.9	27.2	60.7	–	–
Cyprus	209.8	96.3	–	36.0	73.0
Egypt	213.4	203.3	–	–	–
Lebanon	70.7	52.3	8.0	–	–
Turkey	336.5	80.5	102.5	122.0	31.5
Cuba	505.5	285.0	8.0	20.0	192.5
China	94.6	267.4	24.3	–	–
Mexico	45.0	–	–	45.0	–
Honduras	21.0	–	–	–	21.0
Southern hemisphere	907.1	658.9	15.5	112.2	120.5
Argentina	235.5	97.0	–	59.5	50.0
Brazil	138.0	125.5	6.0	–	–
Uruguay	64.0	40.0	9.0	13.0	–
Australia	49.5	44.5	–	–	–
South Africa	416.0	338.0	–	3.0	68.5
Chile	–	–	–	–	–

Source: FAO (1991).

Large industries exist today, primarily where climates are particularly suitable for production and fresh citrus markets could be developed both internally and, more importantly, externally (Albrigo and Behr, 1992). Major fresh fruit exporting countries are shown in Table 1.2. These marketing channels account for about 12% of total world production.

As transportation to other western European countries has improved and populations have increased, the Mediterranean countries have developed a strong fresh fruit export market from their traditional internal use of citrus. Mandarins, lemons and oranges are all important in the total fresh export marketing from the region. Spain has become the major exporting country in the region. Mandarin marketing has remained high because of the development of seedless selections (particularly Clementine and satsuma) and attention to marketing of high-quality fruit. In just 20 years, mandarins increased from only 8% of Spain's production to nearly one-third with increases from 0.16 million tonnes in 1965 to 0.74 million tonnes in 1978 (Del Rivero, 1981) and 1.5 million tonnes in 1989–1991 (FAO, 1991). Improved production techniques, while maintaining quality, have strengthened Spain's marketing position. Some production and price stability should result from the European Community (EC) development.

Conversely, from the late 1970s to the late 1980s, mandarin-type production in and export from the United States has declined (429,000 tonnes total and 18,000 tonnes annual average exports 1976–1981 vs 263,000 tonnes total and 13,000 tonnes annual average exports in 1989–1991. Harvesting without clipping, because of cost, and other poor handling methods in the USA have contributed to quality problems and a loss of market demand. Orlando, Murcott, Dancy, Nova, Robinson and Sunburst have been the primary cultivars in Florida; however, because of their seediness under most growing conditions they are not widely accepted on the world, especially the European, markets.

About four times more oranges than mandarins are produced in the Mediterranean region. Navels and Valencias are the predominant cultivars in Spain and Greece, but in Italy blood oranges such as 'Tarocco', 'Moro' and 'Sanguinello' constitute about 70% of the annual orange production of 1.18 million tonnes (Russo, 1981). 'Biondo Comune' is by far the most popular orange (0.4 million tonnes annually). Italy still produces some bergamots (sour orange hybrids) for perfume essence (Barone *et al.*, 1988), but northern Africa and Paraguay have primarily abandoned their sour orange orchards. Morocco and Egypt have also entered the export market with cultivars similar to those of Spain. Navels and 'Valencias' also predominate in western hemisphere production of oranges for fresh fruit markets.

Development of southern hemisphere export citrus industries in South Africa, Australia and South America somewhat followed the example of South Africa. As the population increased in the late 1800s (due to discovery of diamonds and gold in South Africa), a local citrus industry was started. In about 1906, refrigerated shipping was developed and the UK became involved in fresh citrus marketing. This stimulated further development for export (Oberholzer, 1969). Currently, 'Valencia' and navel oranges make up the major portion of the total production in South Africa. Development of the citrus industry in Australia followed a similar time frame, but after a marketing slow-down in the 1970s, exports have increased to expanding industrialized Asian cities such as Singapore, Bangkok and Hong Kong. The primary production is 'Valencia' and navel oranges. Mandarins ('Ellendale'), grapefruits ('Marsh') and lemons ('Eureka' and 'Villafranca') are produced in near equal amounts and make up one-sixth of the total orange production of about 600,000 tonnes (Gallasch and Ainsworth, 1988). Recent production and planting reflect some increase in the industry.

Many tropical countries in the Americas are increasing planting and should have higher citrus production in the future (FAO, 1991). All tropical and marginally tropical countries of the world produce some citrus, but the major portion of the production is from backyard plantings or small farms and is sold locally. This production is not totally accounted for in available production statistics. Many of the tropical areas have limited production because of severely dry or wet climates and/or disease limitations such as citrus canker, citrus tristeza virus (CTV) or greening (see Chapters 3 and 6). Southeast Asia has been limited in citrus production, primarily to limes and some mandarins, because of these disease constraints. India produces a moderate amount of citrus in spite of these constraints with a balance of orange, mandarin and lime production. Popular cultivars include Mosambi, Blood Red, Pineapple, Hamlin, Jaffa, and Valencia Late oranges and Nagpur Santra and Coorg mandarins. Japan, with its geographically isolated location and cool climate, has developed a historically strong citrus industry, but it is based primarily on the satsuma mandarin because of its citrus canker tolerance and freeze-hardiness. Recent decreasing consumption trends of satsumas in Japan, however, have resulted in a decrease in production (Kitagawa *et al.*, 1988).

World exports of fresh lemons and limes have remained relatively stable from the late 1970s (961,000 tonnes annual average 1976–1981) to the late 1980s (1,002,000 tonnes annual average 1989–1991). Spain, Italy, Turkey, Greece and Egypt produce significant quantities of lemons primarily from the Eureka and Villafranca cultivars. In the western hemisphere, California and Arizona are primary producers of lemons for domestic and export markets, relying on 'Eureka' and 'Lisbon' lemons as their major cultivars. Southern hemisphere countries with subtropical climates, such as Argentina, Uruguay and Chile, have developed a lemon export industry for the off-season in the northern hemisphere with these same cultivars (FAO, 1989). Argentina also produces 'Genova' lemons in addition to 'Eureka' and 'Villafranca'. Mexico is the largest producer and exporter of limes, primarily 'Mexican', in the world. Limes are also a major citrus type grown in most tropical countries in the Americas and Asia.

Grapefruit exports have increased primarily due to increased consumption in Japan, Western Europe, Pacific Rim countries (eastern Asia) and, until the recent economic downturn (1992), Eastern Europe. Major exporters have been the USA (Florida), Cuba, Israel, Argentina, South Africa and Honduras (Albrigo and Behr, 1992). In recent years, production has shifted away from 'Marsh' (white-fleshed) to pink or red cultivars, which are more popular with consumers. The standard 'Ruby Red' is being replaced by cultivars with deeper red flesh coloration, such as Henderson (Flame), Rio Red and Star Ruby.

Table 1.3. Citrus utilized for processing (average of 1989–1990 and 1990–1991 seasons). All values are expressed in 1000 × tonnes. (In some cases values for minor unlisted citrus-growing areas may have been included in the totals.)

Location	Total	Oranges	Mandarins	Lemons/limes	Grapefruits
World	22,356.4	18,613.6	985.3	1,226.9	1,531.0
Northern hemisphere	11,441.4	8,332.6	966.3	792.5	1,350.0
United States	6,908.5	5,628.0	109.5	245.0	926.0
Mediterranean region	3,232.7	2,121.5	396.8	408.0	306.5
Greece	243.4	188.0	9.4	43.0	3.0
Italy	1,210.0	910.0	50.0	250.0	–
Spain	501.5	206.5	223.0	70.0	2.0
Israel	880.3	491.0	89.4	21.5	278.5
Morocco	183.5	168.0	11.5	–	2.0
Cyprus	50.0	17.0	–	13.0	20.0
Egypt	128.0	122.0	5.0	–	–
Turkey	72.0	38.0	17.0	15.0	2.0
Japan	443.0	–	443.0	–	–
Cuba	0	0	–	–	7.0
Mexico	0	0	–	139.5	32.5
China	0	0	–	–	10.0
Southern hemisphere	10,915.4	10,028.0	19.0	434.4	181.0
Argentina	586.0	187.5	13.5	290.0	95.0
Brazil	9,687.0	9,570.0	–	–	21.5
Uruguay	16.0	16.0	–	–	–
Australia	314.0	268.5	5.5	21.0	19.0
South Africa	310.5	239.0	–	26.0	45.5

Source: FAO (1991).

The major citrus-producing countries shown in Table 1.1 are also the major fresh fruit exporters (Table 1.2), although noticeable exceptions are Brazil (primarily São Paulo state) and to a lesser extent, the United States (Florida), where industries are more heavily based on citrus production for processing. The exceptionally large industries in these two locations are the result of their climatic and industrial capacity to produce high-quality, processed, frozen concentrate orange juice (FCOJ) (Table 1.3). While only 33% of all citrus worldwide is processed, 83% of this is oranges with 82% of the orange processing occurring in São Paulo, Brazil and Florida, USA. Florida produces juice primarily from the early, mid and late season cultivars, Hamlin, Pineapple and Valencia (Florida Agricultural Statistics Department, 1990), while in Brazil Hamlin, Pera, Valencia and Natal are the primary cultivars used (Amaro, 1984). 'Pera' accounts for over 50% and 'Natal' 25% of the trees in São Paulo.

The development of a large processing industry occurred because of the invention of FCOJ in Florida in 1948, selection of orange cultivars with high yields, juice content and percentage soluble solids and the continued development of sophisticated industrial infrastructure to meet the world demand for this product (Albrigo and Behr, 1992). In the 1980s, many smaller producing countries in the Americas have begun to develop or expand their orange production with addition of processing facilities for FCOJ. Most lowland tropical locations produce fruit with lower percentage soluble solids, very low acidity and poorer colour than that produced in marginally subtropical regions such as Florida and São Paulo (see Chapters 3 and 7).

Major importing countries of fresh citrus fruit and citrus products include most EC countries, Sweden, Switzerland, Japan, USA, Canada and South Korea. Significant imports are also purchased by Saudi Arabia, Singapore, Israel, Australia and several other Pacific Rim countries besides South Korea and Singapore. Major importing countries that are also producing countries are the USA, Japan, Australia and Israel. Several Eastern European countries imported significant amounts of citrus, although these imports decreased after 1991 (FAO, 1991) due to political restructuring of these countries.

USES OF CITRUS FRUITS

Fresh citrus has been appreciated to the extent that it was carried from its area of origin and cultivated from before the fifth century BC. Early sailors recognized that citrus fruits were important for prevention of scurvy which results from a lack of vitamin C in the diet. Hence, the origin of the term 'limey' for British sailors. The Italians grew citrus in glasshouses (orangeries) perhaps from Roman times (Webber et al., 1967). From the area of origin, citrus was collected, distributed and appreciated by the Arabians, Italians, Spanish and Portuguese. Export of fresh citrus expanded as shipping became available. Sweet oranges, lemons and grapefruit are exported considerable distances by ship (e.g. Florida to Japan), while mandarins are currently shipped smaller distances for regional markets, as is the case in the Mediterranean region. Some long-distance shipping of mandarin fruit does occur from South Africa to Europe, however. Generally mandarin fruits are more easily damaged than oranges or grapefruit during storage and shipping.

Processing of citrus fruit may have begun with the use of citrus peel oils in perfumes from selected cultivars of mandarins by the Chinese in AD 300 and later the use of some sour orange hybrids called bergamots for this same purpose starting in the early 1700s. The first edible processing product may have been orange marmalade prepared from sour orange or candied peel made from citrons. Kumquats and calamondins have been sweet-pickled for many years. Modern processing began with the use of canned sections prepared by hand. Later, machines were developed to peel and separate the sections. Processing greatly expanded with the invention (1948) and promotion (1950s) of FCOJ (Albrigo and Behr, 1992). By-products from juice extraction are important in soft drink production (oils and juice for flavours and juice pulp for cloud), pectin production (peel and rag extracts) and cattle feed production (dried peel pellets). Further details of processing and product handling can be found in Chapter 7.

Citrus production worldwide continues to increase gradually and will probably do so over the next ten years. Current (1993) estimates suggest that citrus production will exceed 80 million tonnes by the year 2000, with major increases occurring in China, Mexico and the US. However, it is likely that supply will exceed demand and prices will stagnate or decline, thus further slowing new development of citrus plantings. Nevertheless, many new markets have become available during the 1980s including former eastern block countries and parts of Asia which will allow for expansion of citrus markets, particularly for fresh citrus products. Moreover, changes in the EEC tariff structure, passage of the North American Free Trade Agreement (NAFTA), and liberalization of import restrictions on citrus in Japan and Korea may expand markets in the future.

2

TAXONOMY, CULTIVARS AND BREEDING

TAXONOMY

Commercial citrus species and related genera are primarily evergreen species of subtropical and tropical origins belonging to the order Geraniales and family Rutaceae. The Rutaceae is one of 12 families in the suborder Geraniineae. Species within the Rutaceae generally have four important characteristics: (i) the presence of oil glands, (ii) the ovary is raised on a floral (nectary) disc, (iii) pellucid dots are present in the leaves, and (iv) fruit have axile placentation (Swingle and Reece, 1967). The family is further subdivided into six subfamilies including the Aurantioideae of which true citrus and related genera are a part. Plants within the Aurantioideae are unusual because the fruit are hesperidium berries (a single enlarged ovary surrounded by a leathery peel) and contain specialized structures, the juice vesicles (sacs). Furthermore, many species contain polyembryonic seeds which may contain both zygotic and nucellar embryos.

The taxonomic situation of tribes, subtribes, genera and species within the Aurantioideae is controversial, complex and sometimes confusing. Citrus and many related genera hybridize readily and have done so in the wild for centuries (Swingle and Reece, 1967). Therefore, there is no clear reproductive separation among species. Moreover, many species reproduce via nucellar embryony. These embryos arise from nucellar tissue and are true-to-type to maternal tissue. Nucellar embryony permits the continued asexual existence of a species or a hybrid.

Taxonomic Systems

The original taxonomic systems of Hooker (1875) and Engler (1896) are artificial systems based on morphological characteristics and the putative origin of a species. Hooker originally proposed that the Aurantioideae consisted of 13 genera with only four species in the genus. Engler later revised this to six species and finally to 11 species in 1931. These taxonomists did not strongly consider the existence and importance of nucellar embryony when developing their classifications, nor did they have access to biotechnological approaches to taxonomy.

During the mid 1900s Swingle (1948) and later Swingle and Reece (1967) developed a taxonomic system based on two tribes, the Clauseneae and the Citreae (Table 2.1). This system classifies citrus and its relatives based on several morphological characteristics and is quite useful and functional from a practical standpoint. The Citreae were further subdivided into three subtribes, the Triphasiinae, Balsamocitrineae and the Citrinae which contains primitive citrus relatives and the true citrus group of six genera including *Citrus*, *Poncirus*, *Eremocitrus*, *Microcitrus*, *Fortunella* and *Clymenia*. Swingle's system further divides *Citrus* into 16 species. A description of each of the six true citrus genera is given below based on the taxonomic system of Swingle.

The genus *Fortunella* (kumquat), named after English plantsman Robert Fortune, includes four species of small trees and shrubs. Leaves are unifoliate and have distinctive silver coloration on their underside. Flowers are borne singly or in clusters in the leaf axils; there are typically five sepals, five white petals, 16–20 stamens, and 3–7 carpels. Kumquats flower much later in the season than other true *Citrus* species and are quite freeze-hardy. Fruit are small, ranging in shape from round ('Meiwa' and 'Marumi') to ovate ('Nagami'). Fruit are served fresh or candied and differ from other true species in that the entire fruit including the peel may be eaten.

Poncirus consists of two trifoliolate species, *Poncirus trifoliata* and *Poncirus polyandra*, and is characterized by small trees with trifoliate, deciduous leaves. The three leaflets are of similar size and oval. *Poncirus* is the only true citrus species that is deciduous and thus is extremely freeze-hardy when fully acclimated. It is native to northern China. There are two distinct types of twigs, those with internodes longer than the petiole and those consisting of foliage spurs which develop from dormant buds and have short internodes. Unlike other true *Citrus* species, bud scales are very pronounced and pubescent, producing long, sharp thorns in the leaf axils throughout the tree's lifetime. Flowers have five sepals, five petals and numerous stamens that curve outward from the base of the ovary. The fruit consists of 8–13 fused carpels; it is pubescent, has a very bitter taste due to the presence of ponciridin and is not palatable. Seeds are plump with a smooth exterior and often contain many embryos. *Poncirus* trees are used as rootstocks but also produce an excellent protective hedge due to their extreme thorniness.

Eremocitrus is characterized by a single species of xerophytic trees native only to Australia and has long been geographically isolated from other true *Citrus* genera except *Microcitrus*. The growth habit is spreading due to presence of long, drooping branches. *Eremocitrus* trees are well-adapted to their xerophytic environment. Leaves are unifoliate, greyish green, thick,

Table 2.1. List of tribes, subtribes, subtribal groups and genera of the orange subfamily, Aurantioideae.

Tribe I. *Clauseneae*: Very remote and remote citroid fruit trees (3 subtribes, 5 genera, 79 species, 20 varieties)

Subtribe A. *Micromelinae*: Very remote citroid fruit trees (1 genus, 9 species, 4 varieties)

1. *Micromelum*: (9 species, 4 varieties)

Subtribe B. *Clauseninae*: Remote citroid fruit trees (3 genera, 69 species, 16 varieties)

2. *Glycosmis*: (35 species)
3. *Clausena*: (23 species, 4 varieties)

Subtribe C. *Merrilliinae*: Large-fruited remote citroid fruit trees (1 genus, 1 species)

4. *Merrillia*: (1 species)

Tribe II. *Citreae*: Citrus and citroid fruit trees (3 subtribes, 9 subtribal groups, 28 genera, 124 species, 18 varieties)

Subtribe A. *Triphasiinae*: Minor citroid fruit trees (3 subtribal groups, 8 genera, 46 species, 3 varieties)

    A. *Wenzelia* group (4 genera, 15 species, 1 variety)

5. *Wenzelia*: (9 species, 1 variety)
6. *Monanthocitrus*: (1 species)
7. *Oxanthera*: (4 species)
8. *Merope*: (1 species)

    B. *Triphasia* group (2 genera, 4 species)

9. *Triphasia*: (3 species)
10. *Pamburus*: (1 species)

    C. *Luvunga* group (2 genera, 27 species, 2 varieties)

11. *Luvunga*: (12 species)
12. *Paramignya*: (15 species, 2 varieties)

Subtribe B. *Citrinae*: Citrus fruit trees (3 subtribal groups, 13 genera, 65 species, 15 varieties)

    A. Primitive citrus fruit trees (5 genera, 14 species)

13. *Severinia*: (6 species)
14. *Pleiospermium*: (5 species)
15. *Burkillanthus*: (1 species)
16. *Limnocitrus*: (1 species)
17. *Hesperethusa*: (1 species)

18. *Citropsis*: (11 species, 1 variety)
19. *Atalantia*: (11 species, 3 varieties)
20. *Fortunella*: (4 species, 1 variety)
21. *Eremocitrus*: (1 species)
22. *Poncirus*: (1 species, 1 variety)
23. *Clymenia*: (1 species)
24. *Microcitrus*: (6 species, 1 variety)
25. *Citrus*: (16 species, 8 varieties)

   B. Near-citrus fruit trees (2 genera, 22 species, 4 varieties)

   C. True citrus fruit trees (6 genera, 29 species, 11 varieties)

Subtribe C. *Balsamocitrinae*: Hard-shelled citroid fruit trees (3 subtribal groups, 7 genera, 13 species)

26. *Swinglea*: (1 species)
27. *Aegle*: (1 species)
28. *Afraegle*: (4 species)
29. *Aeglopsis*: (2 species)
30. *Balsamocitrus*: (1 species)
31. *Feronia*: (1 species)
32. *Feroniella*: (3 species)

   A. Tabog group (1 genus, 1 species)
   B. Bael-fruit group (4 genera, 8 species)

   C. Wood-apple group (2 genera, 4 species)

Totals: 2 tribes, 6 subtribes, 9 subtribal groups, 33 genera, 203 species, 38 varieties.

Source: Swingle (1948).

leathery and lanceolate. Stomata are sunken probably as an adaptive mechanism to reduce transpiration under arid growing conditions. Moreover, *Eremocitrus* appears to be freeze-hardy (Barrett, 1985). Flowers are borne singly or in small groups in the leaf axils, have 3–5 sepals and petals, and 15–20 unfused stamens. The ovary consists of 3–5 fused carpels with two ovules per carpel. The fruit are oval to pyriform in shape with a 'typical' fleshy peel found in all true citrus species. Seeds are small with hard, wrinkled seed coats. *Eremocitrus* is not of commercial importance but has been used in citrus breeding programmes to enhance drought tolerance and freeze-hardiness of the progeny. It has also been used as an interstock.

Microcitrus is indigenous to the northeastern rain-forest of Australia and also has been geographically isolated from other true *Citrus* genera for millions of years. The genus consists of six species of moderate-sized trees. Leaves are unifoliate dimorphic, changing from greatly reduced cataphylls on seedlings to small elliptical to ovate leathery leaves on mature plants. The petiole is small and the upper leaf surface is pubescent. New foliage often has a characteristic purplish tint. Flowers are borne in leaf axils; they are small with 4–5 sepals and petals, 10–20 fused stamens, and 4–8 carpels containing 4–8 ovules per carpel. Petals are white, also having purplish coloration at their base. Fruit are small and roundish, and contain considerable amounts of acidic oil droplets. Seeds are small and ovate in shape. Some species of *Microcitrus australasica* are resistant to burrowing nematode (*Radophilus citrophilus*) and phytophthora fungus (Barrett, 1985). Trees are not nearly as freeze-hardy as *Eremocitrus*. The prefix 'micro' very accurately describes the general size of the leaves, flowers and fruit relative to other true *Citrus* genera. *Microcitrus* has not been of much commercial importance but has been used as an interstock and in rootstock breeding on a limited basis.

Clymenia was also geographically isolated from other true *Citrus* genera millions of years ago and thus differs in a number of ways from the other genera in the true *Citrus* group. It is characterized by a single species of small trees with unifoliate acuminate leaves tapering into a very short petiole. Branches are thornless and subangular when young, changing to cylindrical as they mature. Flowers are borne singly in leaf axils on short, stout pedicels; they have five sepals and petals which contain numerous oil glands. There are 10–20 times more stamens than petals which are fused at the base, but free at the top. The ovary consists of 14–16 fused carpels containing many ovules per carpel. The style is shorter and wider than that of other true *Citrus* species and the stigma is large and flattened. The fruit of Clymenia are ovoid, thin-peeled, contain numerous oil glands and many small seeds. Clymenia is certainly the most unusual and unique of the true *Citrus* group but is not of commercial importance. It has been used as an interstock on occasion.

The genus *Citrus* consists of 16 species of moderate to large evergreen trees based on the taxonomic scheme of Swingle. Tree shape varies from upright in some mandarins to spreading in grapefruit and satsuma mandarins. Branches are angular when young, changing to cylindrical as they mature. Thorns are numerous in leaf axils of young trees or some species like lemons and limes; however, they become less prominent as trees mature. Leaves are unifoliate with laminae ranging in size from very large (grapefruit and pummelos), to moderate (oranges and lemons) to small (most mandarins). Petiole size also varies with species usually in the same manner as leaf size. Flowers are borne singly or in groups in leaf axils and may be either perfect or staminate. Flowers generally have 4–5 sepals, 4–8 petals, 20–40 fused stamens and 8–15 fused carpels usually containing 4–8 ovules in seedy cultivars. The style is longer than the ovary for perfect flowers, but tends to be shorter for staminate flowers. The fruit consists of a single ovary of 8–15 fused carpels (segments) surrounded by a leathery peel. Fruit shape varies from spheroid (oranges) to oblate (grapefruit and mandarins) to prolate (lemons and limes). Juice vesicles have distinct stalks which attach them to the segment walls. Segments are separated by white endocarp tissue called the albedo. The peel contains numerous oil glands and varies in colour from green–yellow (lemons, limes, grapefruit) to orange and reddish orange (oranges) to deep orange to red–orange (mandarins). Seeds are obovoid to roundish in shape and contain one to many embryos. Cotyledon colour ranges from white (oranges and grapefruit) to green (mandarins).

A contemporary of Swingle, T. Tanaka, believed that the differences he observed among many citrus species warranted further division of the genus into 162 species with 13 primary elements (Tanaka,1977). Critics of Tanaka believe that he has included many hybrids that do not warrant species status, while others believe that Swingle has combined too many dissimilar plants into single species. Taxonomists at Cornell University favour a system of identifying some species in the genus using nothomorphs to emphasize the hybrid but sometimes unknown parentage of some citrus species (Bailey and Bailey, 1978). For example, Swingle considered grapefruit (*Citrus paradisi*) a true species; whereas, the Cornell system considers it a natural hybrid thus designated *Citrus × paradisi*. R.W. Hodgson (1967) revised the taxonomic systems of Swingle and Tanaka and with his own observations suggested the use of a system having 36 species. Several other taxonomic systems have been proposed separating *Citrus* into 6–42 species.

Some taxonomists have advocated combining all *Citrus* into a single citrus species because nearly all species readily hybridize. Most taxonomists and certainly most horticulturists view such a combination as impractical and not very utilitarian. Nevertheless, recent studies using chemotaxonomy (Scora and Kumamoto, 1983) and plant morphology (Barrett and Rhodes, 1976; Scora, 1988) suggest that there are only three major affinity groups within *Citrus*, the first consisting of *Citrus medica*, *Citrus aurantifolia* and *Citrus limon* (the *C. medica* group); the second consisting of *Citrus reticulata*, *Citrus sinensis*, *Citrus paradisi*, *Citrus aurantium* and *Citrus jambhiri* (the *C. reticulata* group); and the third consisting of *Citrus maxima* (the *C. maxima* group). A fourth species, *Citrus*

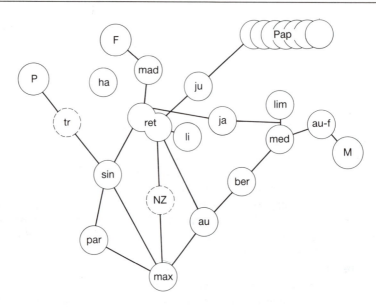

Fig. 2.1. Citrus biotype interrelationships. P = *Poncirus*; F = *Fortunella*; M = *Microcitrus*; Pap = Papedas; ha = *C. halimii*; mad = *C. madurensis* = calamondin; ret = *C. reticulata* = mandarin; sin = *C. sinensis* = sweet orange; max = *C. maxima* = pummelo, shaddock; tr = 'Troyer' citrange; NZ = 'New Zealand' goldfruit; li = *C. limonia* = rangpur; ja = *C. jambhiri* = rough lemon; ju = *C. junos* = yuzu; au = *C. aurantium* = sour orange; ber = *C. bergamia* = bergamot; lim = *C. limon* = lemon; med = *C. medica* = citron; au-f = *C. aurantifolia* = lime. Source: Scora (1988).

halimii, also exists but is not of commercial importance. These affinity groups were derived using biochemical data on oxidase browning and the relative abundance of enzymes, terpenoids, flavenoids, etc. from the various species, as well as morphological and anatomical characteristics such as the degree of polyembryony. Therefore, based on these studies, citron (*C. medica*), shaddock (*C. maxima*) and mandarin (*C. reticulata*) are the only true commercially important citrus species. Sweet oranges appear to be hybrids of mandarins and *C. maxima*, grapefruit of sweet orange and *C. maxima*, lemon of *C. medica* and rough lemon or *C. maxima*, and lime possibly of *C. medica*, *C. maxima* and *Microcitrus*. Possible phylogenetic alliances among species are given in Fig. 2.1 (Scora, 1988).

Further support for this classification comes from molecular biology studies using isozyme analyses and restriction fragment length polymorphisms (RFLPs) (Roose,1988). Isozymes are enzymes with similar biochemical functions in *in vitro* assays but have differing structures. Since their structure is specified by genes, isozyme analysis may indicate genetic relatedness among species. Isozymes are isolated from plant tissues and separated using electrophoresis or isoelectric focusing, two techniques that separate the enzymes based on electrical charge. The presence of the enzymes should be regulated by gene loci whose inheritance pattern is clearly understood to make valid comparisons among species. Similarities and differences are assessed on gels containing samples from various taxa.

Although isozyme analysis is a proven and fairly reliable method of categorizing citrus species, the method is not infallible. The major limitation of isozyme analysis is the relatively small number of isozymes identified, and therefore genes that can be tested (Roose, 1988). Therefore, during the 1980s, techniques using RFLPs were developed to separate species. Theoretically, RFLPs are more accurate indicators of true affinities among species because they can detect differences at the DNA level. RFLPs reflect the locations at which DNA is cleaved by restriction enzymes. The DNA fragments are then separated using electrophoresis and bound to a membrane. The fragments are then hybridized using a DNA probe of known composition. Hybridization occurs in regions where the DNA of the probe and RFLP are homologous. The type of fragments present are then observed using autoradiographs of radioisotopes attached to the probe DNA. The patterns created give insights into the genetic similarities among species. This technique is more powerful than isozyme analysis because some common DNA sequences appear that do not directly affect enzyme synthesis. However, the technique is more difficult and costly than isozyme analysis. As with isozymes, RFLP analysis will indicate if two individual samples are different but cannot indicate positively that the samples are identical because mutations may occur that are not detected in the analysis (Roose, 1988).

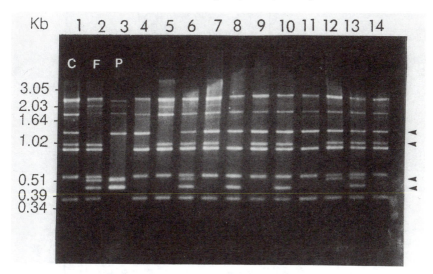

Fig. 2.2. Agarose gel profiles of random amplified DNA products (RADP) from original parents: *C. grandis* (C) and *P. trifoliata* (P), F$_1$ hybrid (F) and some *Citrus* × *Poncirus* backcrosses to *Citrus* (lanes 4–14). Arrowheads indicate polymorphic lanes. Kb = kilobases. Source: G.A. Moore, University of Florida, Gainesville, USA (unpublished).

Table 2.2. A comparison of taxonomic systems for commercially important citrus scion species.

	Scientific name			
Common name	Biochemical/genetic (Scora)	Swingle	Hodgson	Tanaka	Bailey and Bailey
Sweet orange	C. reticulata	C. sinensis	C. sinensis	C. sinensis	C. sinensis
Grapefruit	C. reticulata	C. paradisi	C. paradisi	C. paradisi	C. × paradisi
Lemon	C. medica	C. limon	C. limon	C. limon	C. limon
Lime	C. medica	C. aurantifolia	C. aurantifolia	C. latifolia	C. aurantifolia
Pummelo	C. maxima	C. grandis	C. grandis	C. grandis	C. maxima
Mandarins					
Satsuma	C. reticulata	C. reticulata	C. unshiu	C. unshiu	C. reticulata
Ponkan	C. reticulata	C. reticulata	C. reticulata	C. reticulata	C. reticulata
Dancy	C. reticulata	C. reticulata	C. reticulata	C. tangerina	C. reticulata
Clementine	C. reticulata	C. reticulata	C. reticulata	C. clementina	C. reticulata
Willowleaf	C. reticulata	C. reticulata	C. deliciosa	C. deliciosa	C. reticulata
King	C. reticulata	C. reticulata	C. nobilis	C. nobilis	C. × nobilis
Temple	C. reticulata	C. reticulata	C. temple	C. temple	C. × tangor

Use of RFLPs in citrus taxonomic studies has its disadvantages. The technique is labour intensive, requiring construction of a complementary DNA (cDNA) or genomic library, extraction of relatively large amounts of DNA and the use of ^{32}P and nylon membranes which can be expensive. An alternative method to RFLPs is the random amplified polymorphic DNA (RAPD) method of generating genetic markers (Williams *et al.*, 1992). The technique requires no cDNA clones, uses less DNA than RFLPs and does not use ^{32}P or Southern blots, which are time consuming to run. Basically, the technique involves annealing a single short piece of primer DNA to the template DNA. Polymorphism between species being compared is related to a specific DNA fragment being amplified in one species and not the other. The amplified DNA of each species to be compared is then separated using gel electrophoresis and the different species are compared for regions of similarities and differences. The RAPD method is gaining widespread interest and is likely to replace the RFLP method in the future for much of the genetic mapping in citrus. A typical RAPD gel profile is shown in Fig. 2.2. Note the differences and similarities between bands of *Citrus* and *Poncirus*, their F_1 hybrid and the backcross populations (*Citrus* $\times$ F_1). For example, the second band at 0.51 kb is present in the F_1 and for backcrosses 6, 12 and 13 but not for the other hybrids, or for *Citrus*. The variation in inheritance patterns are polymorphic regions.

Despite the strong biochemical and molecular evidence for only three major affinity groups, there are also practical considerations in any taxo-nomic system. From a commercial and horticultural standpoint, the major species are most conveniently separated based on the system of Hodgson (1968). Hodgson's system, which is predicated on Swingle's, establishes five commercially important species within *Citrus*, the mandarins (*C. reticulata*), sweet oranges (*C. sinensis*), grapefruit (*C. paradisi*), lemons (*C. limon*) and limes (*C. aurantifolia*). These groups are readily identifiable to citriculturists, citrus brokers and consumers compared to the complex system of Tanaka discussed earlier. A comparison of these major taxonomic systems for some commercially important citrus species is presented in Table 2.2.

Hybrids of Citrus and Related Genera

The fact that citrus and related genera readily hybridize has created an interesting group of unusual plant forms and names. The existence of intergeneric hybrids is common in citrus and related genera but unusual in the plant kingdom, further suggesting the lack of clearly defined species as discussed previously. Swingle developed an elaborate system for naming hybrids, some of the most common of which are given in Fig. 2.3 (Jackson, 1991). Of this large and diverse group, the citranges (*C. sinensis* $\times$ *P. trifoliata*) and citrumelos (*C. paradisi* $\times$ *P. trifoliata*) are of greatest commercial importance primarily as rootstocks (Chapter 4). Most of the original crosses

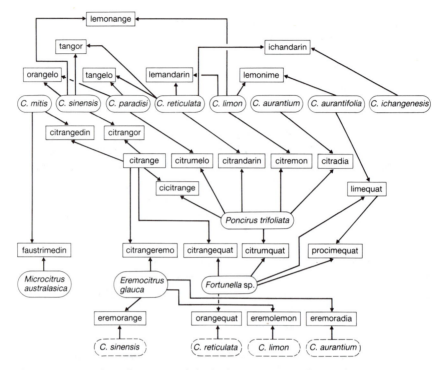

Fig. 2.3. Interspecific and intergeneric hybrids of citrus. Source: Jackson (1991).

were made by W.T. Swingle following the devastating 1894–1895 freezes in Florida. Swingle's intention was to incorporate the freeze-hardiness of *P. trifoliata* into the sweet orange without introducing ponciridin, a bitter principle present in *P. trifoliata* that makes its fruit unpalatable. Unfortunately, while citranges and citrumelos have varying degrees of freeze-hardiness intermediate between the parents, the fruit remain unpalatable. Backcrossing F_1 hybrids improves the edibility of the fruit but also reduces freeze-hardiness. Scions propagated on citrumelos generally have greater freeze-hardiness than those on citranges but this factor may be more related to the vigour induced by the rootstock than to genetic differences in hardiness (see Chapter 5 for further information on freeze-hardiness).

Several other man-made and natural hybrids have achieved limited commercial importance. 'Orlando' and 'Minneola' tangelos (*C. reticulata* × *C. paradisi*) are grown in many areas of the world. In addition, several man-made hybrids have been developed using 'Clementine' mandarin, 'Temple' orange or pummelo as the seed parent because they produce only a single zygotic embryo, thus eliminating problems with nucellar progeny. Of these 'Robinson' and 'Nova' ('Clementine' × 'Orlando') and 'Sunburst' ('Robinson' × 'Osceola') have attained limited commercial importance.

The diversity available in citrus and related genera provides tremendous potential for developing hybrids with desirable characteristics. For example, the inherent freeze- and drought-hardiness traits of *Eremocitrus glauca*, a xerophyte indigenous to Australia, are being incorporated into edible citrus. As with most woody perennial species with protracted juvenility periods, progress in developing new hybrids is slow.

Major Taxonomic Groups within Citrus

On a worldwide basis there are five citrus groups that are of economic significance: sweet oranges (*C. sinensis* [L.] Osb.), mandarins (*C. reticulata* Blanco and *Citrus unshiu* Marc.), grapefruit (*C. paradisi* Macf.), lemons (*C. limon* Burm. f.), and limes (*C. aurantifolia* L.). Kumquats (*Fortunella* sp.) are also grown to a limited extent and the shaddock or pummelo (*C. maxima* L.) is of economic importance in much of southeast Asia and China but not on a worldwide basis. Characteristics of these major species will be discussed in the following sections.

Sweet oranges

Sweet orange (*C. sinensis* [L.] Osb.) possibly originated in northeastern India and central China. Sweet orange is the most widely distributed and enjoys the greatest production of all commercial citrus species (see Chapter 1). Its moderate freeze-hardiness, adaptability to a wide range of climatic conditions and wide range of cultivars make it very adaptable to many growing regions. Sweet orange fruit are in general low to moderate in acids and moderate to high in per cent soluble solids. Therefore, the flavour of fresh fruit and juice products appeals to a broad range of people worldwide.

Sweet oranges may be separated into four groups based on fruit morphological characteristics, chemical constituents, and for convenience: (i) the common or round oranges, (ii) the navel oranges, (iii) the pigmented (blood) oranges and (iv) the acidless oranges. The round oranges are most important commercially and represent a major portion of sweet orange hectarage worldwide. Navel oranges are the second most widely planted group while blood orange plantings are limited primarily to areas with Mediterranean-type climates. Acidless oranges are planted primarily for backyard use and are not of importance commercially.

Sweet oranges are also grouped based on season of maturity, viz. early, mid or late season. In a subtropical climate, early season cultivars generally reach maturity within 6–9 months after full bloom, midseason 9–12 months and late season greater than 12 months. These maturation rates, of course, depend on growing conditions (see Chapter 3) and the factors chosen to define fruit maturity.

Sweet oranges may also be classified based on seediness. Cultivars range in seediness from commercially seedless (0–8 seeds), to moderately seedy (9–15 seeds), to very seedy (>15 seeds). Although many seedy cultivars are used for local consumption, most important commercial cultivars such as Valencia, Pera and navel oranges are commercially seedless. The characteristics of the major round orange cultivars are discussed in the following sections. The cultivars described below by no means represent the large number of local selections that are planted worldwide; however, this group represents a very high percentage of the total production worldwide.

Round oranges

The 'Hamlin' orange originated as a chance seedling near DeLand, Florida, in 1879. Yields for mature 'Hamlin' orchards may average 60–80 tonnes ha^{-1} with some orchards producing over 100 tonnes ha^{-1}. Primarily for this reason 'Hamlin' will continue to be widely planted, particularly in southern flatwoods areas of Florida and some areas of São Paulo state in Brazil. Some selections of 'Hamlin' in Brazil, however, are extremely prone to premature fruit drop. 'Hamlin' is used primarily for processing into juice, but also is marketed as an early season fresh fruit. 'Hamlin' often has heavy early fruit drop in Florida where the cropload is very high but generally fruit hold on the tree well and may be harvested for processing as late as February or March.

'Hamlin' trees grow somewhat upright and symmetrical, and are moderately freeze-hardy. Fruit mature early in the season (September–December [6 months after bloom] in the northern hemisphere) and thus are rarely damaged by freezes or can be rapidly harvested after an early freeze. Fruit are spherical and generally smaller than other sweet orange cultivars. In some seasons fruit size may be limiting for the fresh market. The peel is smooth and thin and when harvested early is susceptible to rind tearing (plugging). Peel colour is generally poor, especially in growing regions with high average annual temperatures. Fruit are susceptible to splitting in some seasons due to the thin peel.

Internal quality of 'Hamlin' fruit is quite poor. Total soluble solids on a fruit basis is the lowest of all major sweet orange cultivars. Moreover, juice must be blended with more highly coloured juices like 'Valencia' to attain minimum processing standards. 'Hamlin' oranges are commercially seedless (fewer than nine seeds per fruit) due to ovule sterility.

'Valencia' was identified and named in Portugal prior to 1865, but most likely is of Chinese origin. Trees from the original selection were shipped to California in 1876 and Florida in 1877 and since then have been widely distributed throughout the world's citrus-growing regions. The original selection in Florida was named 'Hart's Tardiff' due to the late maturity of the fruit. However, the name was later changed to 'Valencia' because the fruits resembled a similar cultivar growing in Valencia, Spain.

'Valencia' is the most important late-season sweet orange in the world. Trees are similar in appearance to most other sweet orange cultivars, with yields varying from moderate (40–50 tonnes ha^{-1}) to high (>60 tonnes ha^{-1}) in some growing regions. Fruit usually matures from February to October in the northern hemisphere and July to September in the southern hemisphere and hold quite well on the tree without serious loss of fruit quality. Fruit held late, however, may regreen on the tree in some areas making them less acceptable for the fresh market. Alternate bearing may be induced in areas where fruit are held late (over 18 months from bloom) due to a reduction in flower bud production and fruit set in the subsequent season.

'Valencia' fruit are of medium size, spherical to oblong, and commercially seedless (fewer than nine seeds per fruit). Fruit quality is excellent, primarily due to the development of deep orange peel and juice colour. 'Valencia' juice is in demand by processors for blending with lower quality juices because of its high levels of total soluble solids (TSS) and colour. Several cultivars of Valencia are prized on the fresh fruit market. These include 'Olinda', 'Frost', 'Campbell', 'Midnight' and 'Delta' among others. These cultivars differ primarily in fruit shape, quality and peel thickness as well as in date of maturity. However, all fruit are very similar morphologically and all mature in late-season.

Because of their late maturity 'Valencia' trees have some unusual problems when grown in relatively freeze-prone subtropical regions like Florida or California (USA). Fruit must remain on the tree throughout the coldest periods of the season. As a result, fruit losses due to freezing may be more severe than those of other, earlier-maturing cultivars. In addition, two crops may be present on the tree at the same time; the mature crop which was set in the previous year and the current year's crop. The presence of two crops causes difficulties in hedging and topping operations because some of the current or future crop is removed. Although removal of a portion of the crop is disconcerting to growers, this practice stimulates new vegetative growth and ultimately improves yields and fruit quality. 'Valencia' fruit are also subject to excessive fruit drop and postharvest losses in some areas due to fruit creasing or splitting (see Chapter 7).

'Natal' is also a late-season sweet orange grown in Brazil primarily for processing. The tree is very similar morphologically to 'Valencia'. Fruit are also similar to those of 'Valencia'.

'Pera' is an important and extensively planted mid- to late-season sweet orange grown for the processing and fresh markets in Brazil (Saunt, 1990). Although 'Pera' is not widely disseminated worldwide, its extensive use in Brazil, the world's largest citrus producer, makes it of relatively great economic importance. Pera trees are morphologically similar to Valencia and other sweet orange cultivars being upright, vigorous and densely foliated. Fruit are medium-sized, ovate and contain 5–10 seeds. Fruit quality is moderate, superior to that of 'Hamlin' but inferior to that of 'Valencia', due to lower juice colour score and TSS. Under Brazil's climate, 'Pera' trees produce multiple blooms and crops during the season. The multiple crops make it difficult to harvest at the correct maturity for best processed quality.

'Shamouti' is a mid-season sweet orange which originated near Jaffa, Israel, in 1844 as a bud mutation of 'Beladi' orange (Saunt, 1990). Sometimes it is referred to as the 'Jaffa' orange which had been grown to a limited extent in Florida, although the two cultivars are not synonymous. 'Shamouti' trees have a slightly different leaf and tree morphology, being more upright and having broader laminae than other sweet orange cultivars. The fruit are ovate and the peel often is quite rough and thick when grown under Mediterranean-type climates. Fruit are commercially seedless and the quality of moderate soluble solids, low acidity and good colour is excellent for the fresh market. Yields are moderate to poor in most citrus-growing regions, although in some areas (southern South Africa) 'Shamouti' oranges are quite productive.

The 'Pineapple' orange is an important mid-season cultivar grown primarily in Florida with some plantings in Brazil and the Republic of South Africa. Trees are moderately vigorous and productive; however, preharvest fruit drop may cause yield reductions. Fruit are seedy which is disadvantageous for the fresh market, but internal quality is excellent. 'Pineapple' juice is deep orange in colour and TSS are high making this cultivar very desirable to citrus processors. 'Pineapple' trees have a tendency toward alternate bearing, which is a disadvantage.

Navel oranges

Navel orange fruit differ from those of most other citrus cultivars due to the presence of a distinctive secondary or even tertiary or quaternary fruit (navel) at the stylar end of the fruit. Tree morphology and leaf characteristics, however, are similar to those of other sweet orange cultivars. Fruit of most navel orange cultivars are seedless due to complete pollen and partial ovule sterility, are generally larger than those of other sweet orange cultivars and are grown primarily for the fresh market (Davies, 1986a). The presence of limonin, a compound that when oxidized imparts bitterness to juice, generally limits the use of navel oranges for processing. Recently, methods have been developed to remove limonin from juice. The juice first undergoes ultra-filtration to remove pulp which interferes with limonin removal. The juice then passes through polystyrene resin exchange columns which remove a high percentage of the limonin (Wethern, 1991). However, this process is relatively slow and moderately costly.

Navel oranges are more susceptible to environmental stresses than other sweet orange cultivars (Davies, 1986a). Insufficient soil moisture during initial fruit set and physiological (June or November) drop periods, and high temperatures during bloom and physiological drop cause significant yield reductions. Moreover, navel orange trees are susceptible to fruit drop during the summer and fall in areas like Florida that may further reduce yield by as much as 30% (Lima and Davies, 1984). Production of navel oranges is especially poor in climates with very high average day- and night-time temperatures (see Chapter 3). However, highest quality fruit are produced in Mediterranean-type climates such as Spain and California coastal regions.

'Washington' navel is by far the most widely planted and commercially important navel orange cultivar (Davies, 1986a). Most other cultivars, with the exception of the Australian group, have originated from Washington either as limb sports or as nucellar seedlings. 'Washington' navel was originally selected because of its superior productivity and fruit quality compared with so-called 'Australian' types. 'Washington' navel is commercially important in California, Australia, Florida, Spain, Morocco and South Africa and is planted in most citrus-growing regions of the world. The trend recently, however, is to replace old-line 'Washington' navel trees with improved local or nucellar selections.

'Atwood', 'Fisher' and 'Newhall' originated from the California budwood certification programme and are among the earliest maturing navel orange cultivars. Tree and fruit characteristics for the most part are similar to those of 'Washington' navel with the exception of 'Newhall' which is a more oblong fruit. Lane Late is a promising late-maturing cultivar which is moderately productive and whose fruit hold well on the tree (Davies, 1986a).

Navelina and Navelate are the most important navel cultivars in Spain, a major producer of navel oranges. 'Navelina' is a limb sport of 'Washington' that originated in Riverside, California, in about 1910. As the name implies, it is usually a smaller tree than 'Washington', is earlier maturing, and fruit hold well on the tree. Fruit tend to be more oblong than other navel cultivars except Newhall. 'Navelate' is a late maturing selection that produces a very vigorous tree. It originated as a sport of 'Washington' in Alcanar, Spain, in 1948.

'Summerfield' navel was widely planted in Florida because it is earlier maturing and appears to be well-adapted to Florida's humid subtropical climate. However, currently nucellar selections such as F-56-11 are the most widely disseminated cultivars and a number of other nucellar selections are showing promise because yields are superior to those of 'Washington' navel. 'Marrs' orange, which lacks a distinct secondary fruit, is a limb sport of 'Washington' navel. It is an early-maturing, productive, low-acid selection widely planted in Texas. 'Marrs' is also lower in limonin than other navel orange selections.

'Baianinha', the most widely planted navel selection in Brazil, originated as a limb sport of 'Bahia'. It is generally less vigorous and has smaller primary and secondary fruit than 'Washington'. Thus, it is less prone to physiological fruit drop than 'Washington' or 'Summerfield' navel under humid subtropical climates. Secondary fruit size appears to be inversely related to incidence of fruit drop (Davies, 1986a). 'Baianinha' also appears to be well adapted to hot, arid growing conditions in northern South Africa.

Leng is a major navel orange cultivar in Australia. It is also a limb sport of 'Washington', found near Mildura, Australia, in 1934. 'Leng' fruit generally are smaller and have a thinner peel than 'Washington' fruit (Davies, 1986a). However, 'Leng' is more susceptible to peel disorders and does not ship as well as 'Washington'. Lane Late, an important late-season cultivar, was selected at Curlwaa, New South Wales, Australia, in 1963. It is similar in fruit size to 'Washington' but may be subject to regreening and granulation late in the season.

'Palmer' is a nucellar seedling of 'Washington' that originated near Brenthoek, South Africa, in the 1930s. It has been the most widely planted cultivar in South Africa since the 1970s. Palmer is a vigorous, productive cultivar with fruit that holds well on the tree. 'Robyn' is a promising selection in cooler locations of South Africa where it matures later than 'Palmer' or 'Washington'.

The 'Cara Cara' navel was discovered in Venezuela. It develops red flesh, even in lowland tropical climates, unlike blood oranges, and is being planted to a limited extent in Florida and South America.

Pigmented oranges

Pigmented (blood) oranges are of commercial importance in several Mediterranean countries including Italy, Spain, Morocco, Algeria and Tunisia, but are not widely grown outside this region. When grown under Mediterranean-type climates with hot days and, most importantly, cool nights, fruit develop very deep red flesh colour which may also occur in the peel. Fruit appearance is very appealing and internal quality is usually excellent. However, flesh colour does not attain the same redness in subtropical or tropical areas with high night-time temperatures. The red colour is due to anthocyanin pigment, which also is the primary pigment in tomatoes and apples, rather than to lycopene or carotenoid pigments which predominate in grapefruit and oranges, respectively. The most important cultivars of blood oranges include Tarocco, Sanguinello, Sanguinelli, Moro and Maltaise Sanguine (Saunt, 1990).

Mandarins

The mandarin group comprises numerous species as well as intergeneric and interspecific hybrids which possess several unique characteristics, including the requirement for cross-pollination for some cultivars to achieve commercially acceptable yields. The term 'mandarin' is used throughout most of the major citrus-producing regions including Japan (the major producer), China, Spain and Italy. The term 'tangerine' is used to refer to most mandarin-type citrus in most of the United States and refers to more deeply pigmented mandarins in Australia and China. Mandarins are referred to as soft citrus in South Africa. Mandarins are produced primarily for the fresh fruit market and as segments (Japan, Spain and China), although the deeply coloured juice may also be blended with orange or other juices to improve their colour, or mandarin juice may be sold as single strength.

Satsuma mandarin group

The progenitor of the satsuma group of mandarins (*C. unshiu* Marc.) probably originated in China but was transported centuries ago to Japan, where it has become the major type of citrus planted. The satsuma itself possibly originated on Nagashima Island, Japan, from seeds brought from China (Saunt, 1990). There are over 100 cultivars of satsuma differing from each other primarily in time of maturity but also in fruit shape and internal quality (Saunt, 1990). The satsuma is well-adapted to cool subtropical regions of Japan, Spain, central China, and southern South Africa and has a low heat unit requirement for fruit maturity. Leaves have large laminae and reduced petiole. The tree has a spreading, drooping growth habit distinctively different from other mandarin types. Satsuma foliage and wood are the most freeze-hardy of all commercially grown citrus cultivars, withstanding minimum wood temperatures of –9°C when fully acclimated (Yelenosky, 1985) (see Chapter 5).

Satsuma fruit are moderately large compared with several other mandarins and are oblate to obovate in shape depending on growing conditions. Fruit grown under humid subtropical conditions tend to be large, obovate and have a coarse exterior often with internal colour changes occurring before peel colour changes. Fruit produced under cool conditions tend to be small, oblate and develop a deep orange peel colour. Satsuma fruit have a moderately hollow central axis and are seedless in most instances due to ovule sterility.

The tendency in recent years in Japan, China and Spain has been to select or breed for earlier maturing satsuma cultivars to extend the harvest season since satsumas, like most mandarins, store poorly on the tree due to segment drying and thus have a limited harvest period. The Japanese have selected for earlier maturing satsumas such as 'Miyagoma' wase and 'Okitsu' wase (wase = early maturing) among others which may be marketable as early as September in the northern hemisphere when field grown (Saunt, 1990). Moreover, when grown under artificial conditions in the greenhouse fruit may be harvested as early as May. Similarly, the Chinese have selected early maturing cultivars Xinjin, Gongchuan and Nangan, also to extend the harvest and marketing period. The 'Wenzhou' satsuma ('Mikan') of China and the 'Owari' of Japan are early to mid-season cultivars, maturing from November to December for 'Wenzhou' and October for 'Owari'. 'Owari' is the most widely planted satsuma in Spain, although 'Clausellina', an early maturing bud mutation of 'Owari', is also planted to a limited extent.

Common mandarin group

The common mandarin group (*C. reticulata* Blanco) represents an assemblage of cultivars. Further, some cultivars in this group, viz. Murcott (Honey) and Ellendale tangor, are naturally occurring hybrids which, based on fruit characteristics, are most suitably classified to this group (Saunt, 1990). Common mandarins differ morphologically from satsumas in having a more upright growth habit and in general small flowers and fruit. Laminae are also generally smaller than those of satsuma, and petiole size is also reduced. Fruit, while having typical mandarin characteristics of a hollow central axis, easy segmentation and green cotyledons, are more difficult to peel than satsumas or fruit in the Mediterranean group. However, the peel is more readily separated than that of sweet oranges. The firmer, more adherent peel also improves handling, storage and shipping characteristics compared with satsuma mandarins. The exceptions are 'Ponkan' and 'Dancy' which when grown under humid subtropical or tropical conditions may produce puffy, difficult-to-ship fruit.

The 'Clementine' mandarin probably originated in China and is similar in characteristics to the 'Canton' mandarin of China. It was selected by Father Clement Rodier in Oran, Morocco, in the 1890s and has become the most widely planted and economically important of all mandarins on a worldwide basis (Saunt, 1990). It is particularly well-adapted and widely planted in North Africa and throughout much of the Mediterranean region and parts of South Africa. 'Clementine' trees are densely foliated, moderately large and have consistently high yields. Fruit quality is excellent but attainment of minimum marketable size is sometimes a problem. Fruit are seedless (with the exception of 'Monreal') and thus are highly prized in the European market. 'Clementine' is not as well-adapted to humid subtropical or tropical growing regions as to Mediterranean climates. Several cultivars of Clementine are available differing in time of harvest, yield and fruit size. 'Fina' has been the mainstay of the Spanish industry, producing a high quality mid-season fruit, although fruit size is less than adequate in some years. As with satsuma, earlier cultivars have been selected to extend the harvest season. 'Marisol' and 'Oroval' are earlier maturing selections of 'Fina'; Nules is a later-maturing cultivar which holds well on the tree.

'Dancy' tangerine was discovered near Orange Mills, Florida, in 1857 by Colonel Dancy and was the most widely planted true tangerine cultivar in Florida. This cultivar is similar to several tangerines grown under other names in other parts of the world. 'Dancy' comes true-to-type from seed because of a high rate of nucellar embryony (nearly 100%) and its period of juvenility is much less than that of sweet oranges (4–5 vs 8–13 years under subtropical conditions). 'Dancy', like many tangerines, tends toward alternate bearing and has typical brittle tangerine wood, a combination often resulting in limb breakage. Fruit size of 'Dancy' is often too small for commercial use in 'on' years. Rough lemon rootstock was commonly used in the past to improve size but lead to early segment drying. 'Dancy' has a very high heat requirement and is well-suited to hot, humid areas. Adequate orange peel colour develops even in tropical areas, but is greatly affected by light, the best colour developing in full sunlight. Poor peel colour is often a problem in densely foliated orchards. 'Dancy' may be spot-picked for both size and colour, although this practice is uncommon in the United States because it is too costly. Most 'Dancy' fruit are shipped fresh in November and December in the northern hemisphere. 'Dancy' fruit hold poorly on the tree, tending to dry out or become ricey; thus it is necessary to market 'Dancy' soon after it meets maturity standards (see Chapter 7). 'Dancy' trees are quite freeze-hardy but the fruit, like all citrus fruit, are not. Moreover, the 'Dancy' tree regrows slowly following freeze-injury. 'Dancy' is not used as a pollinizer for tangerine hybrids because it blooms quite late and in some years ('off' years) does not bloom at all. It is slightly susceptible to *Alternaria* brown spot on fruit and foliage. There have been few new plantings of 'Dancy', because of its many problems and the introduction of mandarin-hybrids; however, Dancy is a very profitable cultivar when managed, handled and marketed properly.

The 'Ponkan' also called the 'Nagpur Santra' in India or 'Warnurco' tangerine, and 'Honey' orange (not to be confused with 'Honey' of California or 'Honey' ['Murcott'] in Florida) is of limited importance worldwide. 'Ponkan' fruit are low in juice acidity, but high in quality and early maturing. Fruit are easily damaged during harvesting and packing due to the inherent puffiness of the peel. Tree growth is very upright, and, like other mandarins, 'Ponkan' has a strong tendency toward alternate bearing and limb breakage is a problem in years of heavy crop load.

Mediterranean 'Willowleaf' mandarin

The Mediterranean mandarin (*Citrus deliciosa* Ten.) also originated in China. It is often called the 'Willowleaf' mandarin because its lanceolate leaves resemble those of a willow tree (*Salix*). The tree is compact, densely foliated and, like most mandarins, is freeze-hardy. The fruit is seedy, of small to moderate size and oblate to necked in shape. Fruit mature in mid-season and flavour is moderate to good.

The popularity of the Mediterranean mandarin has declined recently because the tree has a strong tendency toward alternate bearing and the peel tends to become loose and puffy causing extensive fruit damage to occur during shipping. Improved cultivars of 'Clementine', satsuma and citrus hybrids are currently being planted in lieu of 'Willowleaf' in most citrus regions.

Naturally occurring mandarin hybrids

'Temple' orange (or 'Temple' mandarin as it is sometimes called) is a natural hybrid of tangerine and sweet orange (tangor) which originated in Jamaica in the late 1800s. Budwood of 'Temple' was brought to Florida in 1885. 'Temple' is the most widely-produced mid-season (January–March in the northern hemisphere) mandarin-type in Florida and is also grown to a limited extent in other areas such as South Africa. 'Temple' orange has distinctive tree and fruit characteristics that attest to its hybrid origin. The tree has a dense, spreading growth habit unlike that of most true tangerines. Leaves are lanceolate with thin petioles characteristic of true mandarins. Fruit are of excellent quality possessing high TSS and deep orange juice colour. Fruit, flattened at the stylar end, are obovate to slightly subglobose in shape. Fruit are easier to peel and segment than sweet oranges but more difficult to peel than true mandarins. 'Temple' fruit contain 20–30 seeds with white rather than the bright green cotyledons typical of true mandarins. 'Sue Linda' is a seedless selection from Florida. Some commercially seedless 'Temple' trees are also grown in South Africa.

Because 'Temple' produces only zygotic seedlings, no nucellar virus-free selections were available. Consequently, most old line 'Temple' trees carry exocortis, xyloporosis and citrus tristeza virus (CTV). Moreover, 'Temple' trees tend to produce a number of growth flushes throughout the season, predisposing them to aphid feeding and potential transfer of CTV. The production of many growth flushes also makes 'Temple' trees less hardy to freezes than other mandarin types during most seasons. Recently, shoot–tip grafting has been used to produce virus-free 'Temple' selections. 'Temple' is also highly susceptible to the scab fungus.

'Murcott', also known as 'Honey', 'Honey Murcott' or 'Honey' orange, is a mandarin hybrid which originated either from an abandoned United States Department of Agriculture (USDA) nursery sometime around 1916, or as a chance seedling near Miami, Florida. It was first propagated by Charles Murcott Smith. 'Murcott' is probably a tangor (tangerine × sweet orange), rather than a true tangerine. 'Murcott' has many tree and fruit characteristics typical of a true mandarin. Trees are moderately freeze-hardy and vigorous and have a very distinctive upright growth habit. Leaves are lanceolate, but not as much as those of true mandarins like 'Clementine'. Fruit are borne in terminal clusters causing typical lodging of the branches as in other mandarin cultivars. The fruit are moderately sized, seedy and oblate with a semi-hollow central axis, also characteristics of true mandarins. However, fruit are not as easily peeled as true mandarins and the cotyledons are white, not green. The peel is not loose or puffy and fruit ship well. This combination of fruit and tree characteristics certainly supports the contention that 'Murcott' is a mandarin hybrid.

'Murcott' fruit attain commercial maturity from January to March in the northern hemisphere making them the latest maturing of all commercially important mandarin cultivars. Late maturation has the advantage of placing fruit on the market at times of high prices, but also requires that fruit survive the coldest months of the year risking freeze-damage in areas such as Florida. Most 'Murcott' fruit must be harvested for the fresh fruit market because their juice is high in limonin, a bitter flavour component which makes them less desirable for processing. Internal fruit quality is among the best of any mandarin or mandarin hybrid. Peel and juice attain a deep reddish orange colour at optimum maturity and flavour is excellent.

'Murcott' mandarin is very prone to alternate bearing. Large crops of late-maturing fruit in particular may induce severe alternate bearing with considerably reduced cropping the next season. Moreover, 'Murcott' trees in some instances will set such heavy crops that root growth is severely impaired. Several years of heavy crops will actually kill the tree because carbohydrate reserves in the roots become depleted. This problem, known as 'Murcott' collapse, can be controlled by judicious pruning or by crop thinning to balance the crop and lessen the stress on the tree. Fruit splitting and drop reduce yield significantly in some seasons. Higher N and K levels are recommended for 'Murcott' than for sweet oranges, grapefruit or other mandarin cultivars.

Man-made mandarin hybrids (tangelos)

Tangelos are interspecific hybrids of C. *reticulata* (mandarin or tangerine) and C. *paradisi* (grapefruit or pummelo). The most important commercial tangelo cultivars, Orlando and Minneola, arose from a cross between 'Dancy' tangerine and 'Duncan' grapefruit made by Webber and Swingle in Florida in 1897.

Tangelo fruit and tree characteristics are varied and diverse with some selections being similar to grapefruit and others more similar to true mandarins. In general tangelos are quite vigorous, freeze-hardy and produce weakly parthenocarpic fruit requiring cross-pollination or gibberellic acid (GA$_3$) sprays to achieve adequate fruit set and yields. Leaf morphology ranges from broad and cupped to lanceolate. Petiole size also ranges from narrow to broadly winged.

'Orlando' tangelo produces a vigorous, large tree with distinctively broad, cupped leaves with moderate-sized petioles. The leaves and wood are some of the most freeze-hardy of all commercially important citrus cultivars with the exception of satsumas. Fruit are oblate to subglobose ranging from seedless where no cross-pollination occurs to seedy (10–20 seeds) where sufficient numbers of pollinizers (trees) and pollinators (bees) are present. There is generally a positive correlation between fruit size and seed number. Fruit reach maturity in December to January in the northern hemisphere, attaining bright orange peel colour. 'Orlando' juice has a flavour intermediate between that of mandarin and grapefruit, although a distinct grapefruit flavour is sometimes noticeable.

Cross-pollination with a compatible cultivar like Temple orange or Robinson tangerine is recommended for optimum production. Minneola, its sister cultivar, is not cross-compatible with Orlando and should not be used for cross-pollination. Fruit may also be produced parthenocarpically for fresh fruit markets where seedlessness is desirable. 'Orlando' requires higher rates of N fertilization than round oranges and frequently shows N deficiency symptoms during winter in subtropical areas, known as winter chlorosis. Fruit and foliage are susceptible to the *Alternaria* brown spot fungus.

'Minneola' tangelo ('Honey Bell'), although a sister seedling to 'Orlando', is distinctively different from it in tree and fruit characteristics. Trees are very vigorous, large and spreading with large pointed laminae and moderate-sized petioles. 'Minneola' trees are nearly as freeze-hardy as 'Orlando', but are generally less productive in areas with high springtime temperatures. Fruit are large, obovate and have a pronounced neck at the stem end. Fruit, peel and juice colour are deep reddish orange at maturity and juice flavour is excellent. The peel is moderately adherent and finely pebbled. Seed number ranges from 0 to 10–20 depending on the degree of cross-pollination. Fruit of 'Minneola' reach commercial maturity between January and March in the northern hemisphere and July to August in the southern hemisphere. 'Minneola' like 'Orlando' requires cross-pollination with, for example, 'Temple' orange or 'Robinson' tangerine to achieve adequate yields and optimum fruit size. However, fruit can be produced parthenocarpically by planting large solid blocks and using GA_3 postbloom for fresh fruit markets where seedlessness is desirable.

Other man-made mandarin hybrids

The high incidence of nucellar embryony in citrus has caused considerable problems for fruit breeders wishing to develop new citrus cultivars. Consequently, breeders have found it necessary to use seed parents that produce only zygotic progeny in many of their crosses. In 1942, Gardner and Bellows at the USDA research laboratory in Orlando, Florida, crossed 'Orlando' tangelo with 'Clementine' mandarin (zygotic progeny only). The resulting seedlings were planted and screened for desirable characteristics. Four selections were named and released in 1959: 'Robinson', 'Lee', 'Osceola' and 'Nova'. A cross between 'Minneola' tangelo and 'Clementine' was also made in 1942 leading to the release of 'Page' orange in 1963. 'Sunburst' tangerine, resulting from a cross between 'Robinson' and 'Osceola', was released by the USDA in 1982. 'Fallglo', a 'Bower' × 'Temple' cross made in 1962 by P.C. Reese, was selected for further testing in 1972 and released for propagation in 1987. 'Ambersweet', a 1989 release, resulted from a cross between a mandarin hybrid ('Clementine' × 'Orlando') and a mid-season sweet orange

(15-3) made in 1963 by C.J. Hearn and P.C. Reese near Fort Pierce, Florida. A seedling from this cross was selected in 1972. These hybrids are grown to a limited extent worldwide and are produced primarily for the fresh fruit market.

'Robinson' trees are moderately vigorous and similar in morphology to true mandarins. 'Robinson' requires cross-pollination with compatible culti-vars such as Temple or Orlando to produce adequate yields under most growing conditions, although fruit are also produced parthenocarpically. Sometimes the fruit load becomes excessive, causing limb breakage. Leaves and wood of 'Robinson' are very freeze-hardy, but tree vigour is usually decreased due to excessive cropping or 'Robinson' dieback and freeze damage often occurs. Fruit of 'Robinson' generally mature from October to December in the northern hemisphere. Fruit tend to be small to moderate in size and oblate in shape. External and internal colour attain a deep, rich orange hue, although in some years changes in peel colour lag behind internal quality changes causing difficulty in judging maturity. The fruit has typical mandarin characteristics of easily separable peel and segments and a hollow, open centre. The fruit quality of 'Robinson' is excellent. Seed number varies from 0 to more than 20 per fruit depending on the degree of cross-pollination. In general fruit with more seeds are larger than those with 0 to a few seeds. Seeds produce zygotic progeny only. 'Robinson' is prone to limb dieback which reduces yields. Fruit are susceptible to splitting and plugging because of excessively thin peels in some years. Small fruit size is also a problem in some seasons. Fruit generally do not hold well on the tree or ship well.

'Sunburst' tree and fruit characteristics are similar to those of 'Nova' and 'Robinson'. Trees are moderately freeze-hardy and tolerant of citrus snow scale, but the foliage and stems are very susceptible to rust mite damage and some other problems under Florida conditions (Albrigo et al., 1987). Fruit are oblate, highly coloured, and thin-peeled; they mature after 'Robinson' (October–November) in the northern hemisphere but before 'Dancy'. As with other mandarin hybrids, seed number varies with degree of cross-pollination. External and internal colour generally develop together. 'Sunburst' fruit ship better than fruit of 'Robinson' or 'Dancy'. 'Sunburst' requires pollination with cross-compatible cultivars like Temple, Orlando or Nova to produce an acceptable crop. Seeds of 'Sunburst' are polyembryonic.

'Fallglo' is an early-season cultivar which is harvested between late October and November in the northern hemisphere. Growth and leaf characteristics are similar to 'Temple' orange, being densely foliated, moder-ately vigorous and upright but somewhat spreading. Leaves are lanceolate with reduced petioles reminiscent of 'Temple', but are resistant to sour orange scab. 'Fallglo' trees are less freeze-hardy than those of 'Orlando' or 'Sunburst' and may be less suitable for planting in areas where freezes are likely to occur. The oblate fruit of 'Fallglo' are relatively large with a small but distinctive navel and contain 30–40 seeds depending on degree of cross-pollination.

Seeds of 'Fallglo' produce only zygotic progeny. The peel is smooth, contains prominent oil glands and is easily removed. Peel and juice colour are deep orange and fruit quality is comparable with that of 'Temple'. Unlike many other mandarin hybrids, 'Fallglo' seems not to require cross-pollination to achieve optimum production. Further field trials, however, are needed to substantiate this.

'Ambersweet' is an early-season (mid-October to December) hybrid. Tree growth is somewhat upright and moderately vigorous. Trees have intermediate freeze-hardiness. Fruit are moderately large, convex-shaped and tapered at the stem end somewhat similar to 'Temple' orange. Occasionally fruit produce a small navel. Juice colour is dark orange under growing conditions typical of northern Florida. Fruit range from seedless to seedy (15 or more) depending on degree of cross-pollination, although unlike 'Robinson' and 'Sunburst', cross-pollination does not appear necessary for adequate yields. Like 'Robinson', seeds produce zygotic progeny only and most have white cotyledons, although some green ones occur. Juice taste and quality are similar to that of round oranges except with better colour. Recently, regulations were changed to allow full use of this cultivar in frozen concentrated orange juice (see Chapter 7).

Most mandarin hybrids and some mandarins differ from other citrus cultivars in their requirement for cross-pollination to attain commercially acceptable yields. Characteristics of a good pollinizer (tree) cultivar include overlapping bloom periods, adequate flowering, production of saleable fruit, similar freeze-hardiness and cultural requirements, attractiveness to bees (the agents for cross-pollination) and cross- or self-compatibility with the main cultivar. Pollinizer placement varies from orchard to orchard from a 1:3 to 1:1 ratio (pollinizer:main cultivar). In some cases, pollinizers are interplanted within the row every third to fourth tree. This pattern may cause problems during harvesting, however, due to mixing of fruit.

Grapefruit

Grapefruit (*C. paradisi* Macf.), as noted earlier, is probably not a true species but a hybrid of pummelo and sweet orange (Scora, 1988). Grapefruit is one of the few citrus types that originated in the New World, most probably in the West Indies. The pummelo parentage is quite apparent in similarities between fruit and leaf morphology. Grapefruit, unlike pummelo, however, produces nucellar and zygotic embryos, rather than zygotic embryos only.

Grapefruit production and distribution are much more limited in scope than those of sweet oranges or mandarins. The high heat requirement limits production of highest quality fruit to tropical and hot, humid subtropical regions. In contrast grapefruit grown in Mediterranean climates often are high in juice acidity and have a thicker peel and lower juice content than those grown in more tropical or humid subtropical regions. The high juice acid and moderately low TSS levels for grapefruit in general produce fruit less palatable to many people than oranges, and fruit are less often peeled and eaten out-of-hand as are other citrus fruits. Grapefruit consumption is greatest in North America, Europe and Japan; yet, pummelo is far more popular in China and most of southern Asia.

Grapefruit trees are very vigorous, having a spreading-type growth habit reminiscent of their pummelo heritage. The structural framework is very sturdy and capable of supporting the greatest crop loads of any commercial citrus type. Leaves are unifoliately compound and much larger than those of sweet orange, mandarins, lemons or limes, and are similar in appearance, although smaller, to those of pummelo. The lamina is ovate with entire to slightly serrated margins. The petiole is large and winged, often with a distinct cordate shape. Grapefruit generally possess a 2/5 phyllotaxis in contrast to the 3/8 phyllotaxis of sweet orange. Grapefruit flowers are perfect and complete having the same basic structure as that of other citrus species. Flowers are larger than those of other commercially important cultivars with the exception of pummelo. Flowers and thus fruit are often borne in clusters, hence the origin of its name, although fruit are also borne singly particularly in the interior portions of the tree.

Fruit of grapefruit are the largest of any commercial citrus cultivar, again with the exception of pummelo. Fruit diameter ranges from 8 to 14 cm at maturity depending on cultivar, rootstock and growing conditions. Grapefruit may remain on the tree for protracted periods, and are harvested from September to July in the northern hemisphere. However, some fruit drop may occur which reduces yields, and fruit held too late may become granulated or seeds may germinate within the fruit on the tree (vivipary) thus lowering marketability. Grapefruit quality generally improves if it remains on the tree and juice acid levels decrease.

Fruit vary from quite seedy to seedless; the latter are preferred on the fresh market. Seeds are large and plump with polyembryonic, cream-coloured to white cotyledons.

White-fleshed grapefruit

Duncan grapefruit has been the major seedy grapefruit cultivar in the United States; however, production has been declining and few new plantings are being made either in the United States or worldwide. 'Duncan' probably originated as a seedling of the original grapefruit seedlings brought to Florida by Don Phillipe. It was first discovered in the orchard of A.L. Duncan near Safety Harbor in 1875, but was not named until 1892.

Duncan is a typical, vigorous grapefruit cultivar. Consequently, trees are generally set at wider spacing than those used for sweet oranges or mandarins. Fruit are produced terminally in clusters and tend to be larger and higher in total soluble solids than those of 'Marsh'. 'Duncan' is used primarily for processing into sections, juice vesicle slurry and juice, since seediness limits its demand as fresh fruit.

Marsh grapefruit is the most widely planted white-fleshed grapefruit cultivar in the world. It originated as a chance seedling around 1860 near Lakeland, Florida, but was not propagated commercially until the late 1880s by C.M. Marsh, a commercial nurseryman. 'Marsh' trees are very similar in vegetative characteristics to other grapefruit cultivars, but fruit size and quality are generally not as good as those of 'Duncan'. The major reason for the popularity of 'Marsh' is its seedlessness and the fact that it holds well on the tree without significant loss of fruit quality, thus extending the harvest season if desired. Fruit are used for processing, but are grown primarily for the fresh market.

Red-fleshed grapefruit

All currently popular pink- and red-fleshed grapefruit cultivars have arisen as mutations of white-, pink- or red-fleshed cultivars. The general trend since the early 1900s has been to select and propagate cultivars with progressively redder flesh and peel colours. This selection process has occurred due to increasing consumer demand, and consequently higher prices for redder-fleshed cultivars. In general these mutations do not differ in internal quality except in the degree of redness with the exception of 'Star Ruby' which has more coarse-textured flesh than other cultivars. The deeper red colour, however, increases the eye appeal to the consumer and thus increases demand.

There are two major developmental lines giving rise to the most important pink- and red-fleshed grapefruit cultivars (Fig. 2.4). The first is derived from Walters (a white-fleshed cultivar) and includes Foster, Hudson and Star Ruby, and the second line derives from Thompson (Pink Marsh), which includes Redblush, Ruby Red, Henninger, Ray Ruby, Rio Red, Flame (Henderson) and Burgundy (Saunt, 1990).

'Foster', the first of the pink-fleshed mutations, occurred as a limb sport of 'Walters' in the Atwood orchard near Manatee, Florida, in 1907. The pulp is light pink but the juice of 'Foster' is nearly colourless. 'Foster' was not widely planted primarily because it is seedy but it has been the source of other more important red-fleshed cultivars. 'Hudson' originated as a bud-sport of 'Foster' in the 1930s in San Benito, Texas. 'Hudson' has deeper red flesh and peel colour than 'Foster' but is also seedy. In 1959, R.A. Hensz of Texas A & I University irradiated seeds of 'Hudson', planted them and subsequently evaluated the seedlings. In 1970 he selected a very deep red-fleshed seedling which he named and released as 'Star Ruby'. Star Ruby has the deepest red peel, flesh and juice colour of any currently available red grapefruit cultivar. The red coloration is even found in the bark of young branches. The leaf colour of Star Ruby differs from that of other red-fleshed cultivars becoming

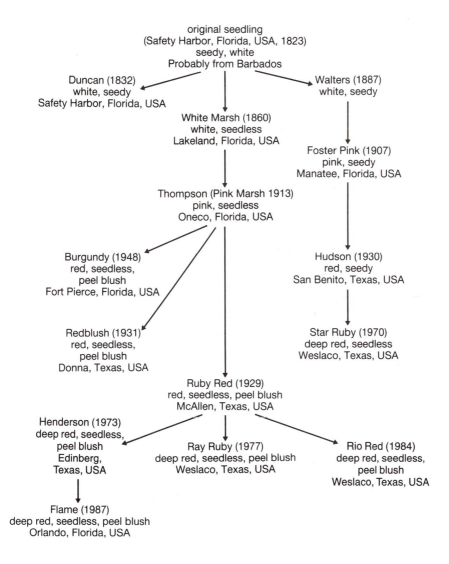

Fig. 2.4. Origin of the most commercially important grapefruit cultivars. Numbers in parentheses indicate date of discovery or release. Adapted from Saunt (1990).

chlorotic, almost white in some orchards. 'Star Ruby' leaves also become chlorotic in response to some herbicide applications and trees are quite susceptible to *Phytophthora* foot rot. However, despite these problems, yields, fruit quality and economic returns have been very good.

Another line of red-fleshed grapefruit originated from 'Marsh' beginning with 'Thompson' ('Pink Marsh'), a limb sport in the orchard of W.B. Thompson in Oneco, Florida (USA), in 1913. 'Thompson' has light pink flesh and peel colour even under optimum conditions, but has the advantage over 'Foster' in being seedless. Tree characteristics of 'Thompson' are identical to those of 'Marsh'. Several of the red-fleshed cultivars are mutations of 'Thompson'. 'Ruby Red' ('Ruby') was discovered as a limb sport of 'Thompson' by A.E. Henninger near McAllen, Texas, in 1929. In 1931 another limb sport of 'Thompson' was selected by J.B. Webb of Donna, Texas, and propagated as 'Redblush' ('Webb Redblush'). These cultivars are nearly identical in fruit and tree characteristics. In both, fruit are more deeply red-pigmented than 'Pink Marsh' and moreover a distinct red peel blush is present.

'Ruby Red' has been the source of three more deeply coloured cultivars in Texas, 'Henderson', 'Ray Ruby' and 'Rio Red'. 'Henderson' was first observed as a limb sport of 'Everhard' grapefruit in 1973 in the Henderson orchard near Edinburg, Texas. It has deeper red flesh and peel colour than 'Ruby Red'. 'Ray Ruby' is also a mutation of 'Ruby Red' with peel colour similar to that of 'Henderson' but with deeper red flesh colour than 'Henderson'. 'Rio Red' originated as a seedling of 'Ruby Red' planted in 1953 and selected because of favourable yield and fruit characteristics in 1959. Budwood of this selection was irradiated and propagated on sour orange rootstock. A mutation was selected in 1976 for its deep red coloration, named and released as 'Rio Red' in 1984. Peel colour is similar to that of 'Ray Ruby', but the flesh colour is redder.

A fourth red-fleshed cultivar, Flame, was released for propagation in 1987 by the USDA in Florida. 'Flame' originated from seeds of 'Henderson' planted in 1973. Peel colour is similar to that of 'Ray Ruby' but internal colour is nearly comparable to that of 'Star Ruby' when grown under Texas conditions. In some areas of Florida trees have shown leaf chlorosis symptoms similar to those of 'Star Ruby', but internal colour seems to be more stable than that of 'Rio Red' under Florida conditions.

An important characteristic of 'Ray Ruby', 'Rio Red', 'Flame' and 'Star Ruby' not found in 'Ruby Red' is that the deep red internal colour persists much later in the harvest season. This extends the marketing period. Variations in colour may occur depending on climatic conditions, although the red colour develops even in hot, lowland tropical regions.

Limes

Lime (*C. aurantifolia* L.) trees most probably originated in tropical areas along the Malay archipelago and as a result of this heritage are the most freeze-sensitive of all commercial citrus species (Yelenosky, 1985). Thus, their distribution is limited to the tropics and warm, humid subtropical regions where minimum temperatures remain above −2 to −3°C. The two major groups include the acid and acidless limes of which the acid limes are of commercial importance. The acid limes are further subdivided into 'Tahiti' ('Persian', 'Bearss') and 'Key' ('West Indian', 'Mexican') limes.

Lime trees are extremely vigorous, possessing an upright to very spreading growth habit. Trees are often quite thorny. Leaf morphology differs between the two major groups. The lamina of the 'Tahiti' lime is large and elliptical to ovate with noticeable serrations near the apex. The 'West Indian' lime lamina is much smaller and ovate to nearly round in shape. The lamina is distinctly serrated around the apical margin. Petioles of both types are very reduced and nearly nonexistent.

Lime flowers are perfect and complete having the same general characteristics as other citrus species in most cases. However, occasionally some 'juvenile' flowers are produced which have reduced or nonexistent gynoecia. Flowers of 'Tahiti' lime, like lemon, are purple-petalled, while those of 'West Indian' lime are white-petalled and smaller than those of 'Tahiti'. Petals of 'West Indian' lime are often extrorse, curving downward from the central axis. Flowering generally occurs in two major peaks seasonally, but also may occur continuously at reduced intensity producing several crops per year.

Fruit of 'Tahiti' limes are very similar morphologically to lemons (prolate spheroid to spherical); those of 'West Indian' lime are spherical and much smaller than those of 'Tahiti' lime. 'Tahiti' lime rarely produces seeds due to near complete ovule sterility. 'West Indian' lime is moderately seedy having 10–15 seeds per fruit.

Lemons

Lemon (*C. limon* Burm. f.) trees probably originated in the eastern Himalayan regions of India, although this conjecture is based only on observations of wild species growing in this region. It is generally accepted that lemon is a hybrid closely related to citron (Scora, 1988). Distribution and major production areas of commercial lemons are limited primarily to semiarid to arid subtropical regions having minimum temperatures greater than −4°C. Lemon trees are more sensitive to low temperatures than other commercial cultivars with the exception of limes (Yelenosky, 1985), and are not well-adapted to humid subtropical or tropical regions due to susceptibility to fungal and algal diseases. In addition, fruit quality, especially peel texture, which becomes coarse, is generally poorer in humid regions than in Mediterranean-type climates.

The lemon tree is extremely vigorous, having an upright, rank and unwieldy growth habit when juvenile which continues when grown in tropical areas, another reason for poor adaptability to tropical climates. As trees mature or when they are grown in Mediterranean climates the growth habit becomes more spreading and manageable. Trees are often quite thorny, although this varies with cultivar, growing conditions and tree age. Generally, young trees are very thorny and leaf morphology is quite variable depending on tree vigour. Laminae are large and ovate with pronounced serrations along the apical margins of leaves developing on vigorous shoots. Laminae become ovate to lanceolate again with serrated margins as shoots mature. Newly developing leaves are purple but as the laminae mature they become green. Petioles are reduced, and nearly nonexistent in some instances.

Lemon flowers are perfect and complete having the same general characteristics as other commercial citrus species. Flower petals, like newly expanding leaves, are purplish white. Flowers are smaller than those of grapefruit but similar in size to mandarin flowers. They are typically borne in clusters. Lemon trees have two major flowering periods in Mediterranean climates but tend to flower continuously throughout the year in cool, coastal regions such as those of California.

Fruit are produced throughout the year in most major growing regions, but depending on cultivar and environmental factors a large part of the crop is usually harvested in either the summer, autumn or winter. Fruit shape varies from nearly spherical in some cultivars to the more common prolate spheroid of most commercial selections. Fruit have a characteristic apical mammila (nipple) at the stylar end. Lemon fruit are high in total acidity (TA) (5–8%) and low in TSS (7–9%) relative to any other commercially important citrus cultivar except acid limes. Cultivars range from moderately seedy to seedless. Seeds are small with smooth seed coats and pointed micropylar ends.

There are three major groups of lemons not including those used as rootstocks: the *Femminello*, Verna (Berna) and Sicilian types. The *Femminello* and Verna types are primarily grown in North Africa and Europe and Sicilian types in the United States and South Africa (Saunt, 1990). The *Femminello* group, *Femminello comune*, *Femminello siracusano* and *Femminello St Teresa*, represents the most widely grown lemon types in Italy. Since the cultivars are everbearing several crops can be harvested throughout the season. The Primofiore crop is harvested from September to November; the Limoni crop from December to May; the Bianchetti crop from April to June and the Verdelli crop from June to October (Saunt, 1990). Tree morphology is characteristic of most lemons in being upright and vigorous. Yields are moderate to high and are quite consistent. *Femminello* accessions are moderately seedy and fruit shape prolate spherical with a moderately sized apical mammila. *Femminello* fruit have moderately high acidity and lower than typical juice content. They are marketed fresh or for their peel oil content.

Verna is the major cultivar of Spain and is very similar morphologically to 'Lisbon' lemon. Unlike the *Femminello* group, 'Verna' trees produce a major crop (February–August) and a second lesser crop (Verdelli, August–October) in the northern hemisphere. Fruit are more elongated and have a more pronounced apical mammilla than *Femminello* but are seedless.

The most important cultivar in the Sicilian group is Eureka, which is also important in California and Australia. It originated as a seedling from Sicily brought to California in 1858. It is also grown in South Africa, Spain and Israel (Saunt, 1990). Tree morphology is slightly different from some other cultivars in having a less densely foliated spreading canopy. Consequently, 'Eureka' is considered to be less frost-hardy in California than 'Lisbon'. Fruit are moderate- to small-sized and ovate with a moderately rounded apical mammilla. Fruit quality is excellent when grown in coastal, Mediterranean-type climates. The peel is smooth and thin and fruit have high juice and acid levels. Fruit usually have fewer than nine seeds and are produced throughout most of the year. Yields of 'Eureka' are generally lower than those of 'Lisbon' in California or 'Fino' in Spain.

'Lisbon', like 'Eureka', is grown primarily outside the Mediterranean region, namely in Australia, California and Argentina, where it is better adapted to the climate. Trees are densely foliated and vary in degree of thorniness. The dense foliation delays radiation losses from the interior canopy more effectively than the less densely foliated 'Eureka'. Thus trees withstand some frosts better than 'Eureka' trees. However, there is probably no inherent physiological difference in freeze-hardiness between Lisbon and other lemon cultivars. Fruit are excellent in quality differing slightly from those of 'Eureka' under some growing conditions. The apical mammilla and areolar furrow are more pronounced on 'Lisbon', but under most conditions fruit of 'Lisbon' are very similar to those of 'Eureka'. Major harvest periods occur in winter and spring. Fruit usually have fewer than nine seeds. 'Lisbon' has replaced 'Eureka' as a major cultivar in California because of superior yields. Moreover, the denser foliage provides some protection from wind and frost damage to the fruit.

Pummelo (shaddock)

Pummelo (*Citrus grandis* [L.] Osb.) is probably native to southern China or the Malay and Indian archipelagos. It is sometimes referred to as a shaddock and is grown to a very limited extent only in a few of the major citrus-producing regions. It is a popular fruit, however, in much of China, and southeast Asia. Although the tree and fruit look similar to grapefruit there are some important differences. Leaves, flowers and fruit are usually the largest of any citrus type, despite some variability. The lamina is ovate with a moderately serrated margin; the petiole is cordate and distinctly winged and separate from the lamina. The young twigs and leaf midribs are often pubescent. Flowers have the same basic structure as other true citrus, but they are the largest of the true citrus species. The fruit shape ranges from obovoid to pyriform with a pronounced flattening of the stylar end in most cultivars. Fruit diameter ranges from 10 to 30 cm, but may be significantly larger than this for some cultivars. The internal quality of pummelos differs from that of grapefruit. The peel is extremely thick but easy to peel and the juice sacs are very pronounced and rubbery in appearance and texture. Generally, the rind and the segment walls are peeled before eating. Pummelo juice is not bitter as that of grapefruit and thus it has a sweet, mild flavour. Seeds are large, plump and produce only zygotic progeny for nearly all cultivars.

There are many cultivars of pummelo but they are generally divided into the Thai, Chinese and Indonesian groups. Fruit in the Thai group are generally smaller than those in the Chinese group (Saunt, 1990). Major cultivars in the Thai group include Chander (pink-fleshed), Kao Panne and Kao Phuang (white-fleshed); those in the Chinese group include Goliath, Mato and Shatinyu (all white-fleshed); those in the Indonesian group include Banpeiyu (white-fleshed) and Djeroek Deleema Kopjar (pink-fleshed). There are also two hybrids between pummelo and grapefruit, 'Melogold' and 'Oroblanco', which have intermediate characteristics between the two and are being grown to a limited extent in California and Israel.

Kumquat

Kumquat (*Fortunella* spp.) originated in northern China and due to several differences in morphology have been placed into the genus *Fortunella* rather than *Citrus* by several taxonomists. Kumquats are produced primarily in China and the Philippines with limited production in other areas of the world. Kumquats are eaten fresh or candied and are unique in that the entire fruit including the peel is generally eaten whole. The tree is moderately vigorous and upright to spreading in its growth habit. Leaves are small with lanceolate to elliptical laminae and greatly reduced petioles. The underside of the lamina has a characteristic silvery colour. Kumquats tend to bloom much later than other commercial cultivars and the leaves and wood are quite freeze-hardy when fully acclimated.

Fruit shape varies with species and cultivar from elliptical ('Nagami', *Fortunella margarita*) to spherical ('Meiwa', *Fortunella crassifolia*; 'Marumi', *Fortunella japonica*) and fruit size ranges from 2 to 3 cm in diameter. Fruit are composed primarily of peel, albedo and flavedo tissue and have a very limited amount of juice. Kumquats differ from other true citrus in having only three to five carpels, each containing one to two seeds. Seeds are small and have green cotyledons.

CITRUS GENETICS AND BREEDING

Citrus and related genera within the true citrus group have two sets of nine chromosomes ($2x = 18$). Chromosomes are small and variable in size and aberrations are moderately common (Raghuvanshi, 1968). Some triploids, tetraploids and hexaploids exist but generally occur at low percentages in the population. Some research suggests that percentage of polyploids is greater in large than in small seeds and may vary with fruit position on the tree.

Selection of new citrus and related cultivars has been occurring for thousands of years beginning in ancient China where superior phenotypes were selected from the wild for cultivation. However, systematic, mission-oriented breeding programmes first began in Florida in 1893 with Swingle and Webber. Since then numerous programmes have been developed worldwide with a variety of objectives. Unfortunately, citrus breeding is difficult and time consuming. Most citrus and related species are very heterozygous and few important traits show single gene inheritance patterns; therefore F_1 hybrids tend to exhibit variability. Furthermore, the common occurrence of nucellar embryony and absence of characteristic morphological marker genes make selection of hybrids difficult, although this situation is improving through the use of isozyme markers. Also, the protracted juvenility period of seedlings in the field (5–15 years) makes citrus breeding a very long-term, costly and land-intensive proposition. Consequently, most cultivars of worldwide importance have resulted from mutation or natural hybridization in the wild. For example, pink and red grapefruit were selected from limb sports of white or other coloured grapefruit and navel oranges are probably limb sports of round sweet oranges. Several earlier maturing satsuma and 'Clementine' selections have also resulted from mutation. Grapefruit and some mandarins like 'Temple' orange or 'Murcott' mandarin are most probably naturally occurring hybrids of oranges and mandarins. Natural hybridization in the wild is common. Many hybrids are perpetuated by nucellar embryony and are contributing factors to the confusing species situation in citrus.

Nucellar embryony

The percentage of nucellar seedlings ranges from zero for zygotic species such as *C. maxima* and cultivars such as Clementine mandarin and Temple orange to virtually 100% for 'Dancy' and 'Kara' mandarins. Most citrus and related species are highly heterozygous and thus produce widely variant sexual (zygotic) progeny. The continued production of weak, noncompetitive zygotic offspring due to inbreeding depression may have led over time to selection for higher levels of nucellar embryony. The tendency to produce nucellar compared with zygotic embryos is inherited simply in a 1:1 ratio in some instances, but deviates from this ratio in other crosses depending on the species involved (Cameron and Soost, 1979).

In the past, degree of nucellar embryony was determined largely by visual ratings. In some instances, nucellar and zygotic embryos are difficult to separate in this way due to the relative lack of distinctive morphological genetic markers especially when hybridizing closely related cultivars or species such as 'Pineapple' and 'Valencia' sweet oranges or when zygotic seedlings result from self-pollination of a relatively homozygous parent. During the 1970s, methods were developed to distinguish between embryo types using isozyme analysis (Torres *et al.*, 1978). Although this method is not foolproof, it is usually more accurate than visual methods. Xiang and Roose (1988) compared the relative degree of nucellar embryony of some selected citrus rootstocks based on isozyme analysis. Among ten rootstock accessions the percentage nucellar embryony ranged from 49.4 for 'Yuma' citrange to 94.5 for 'Indio' rough lemon. Moreover, percentage nucellar embryony varied for different seed lots from within the same species, e.g. 84.8 vs 94.0% for CPB4475 citrumelo. Selections with high percentages of nucellar embryos are important for the production of uniform, true-to-type citrus rootstocks (see Chapter 4). Moreover, nucellar trees are usually virus-free because most viruses (viroids) are not translocated to the developing embryo. There is evidence, however, that psorosis has been transmitted in seed of 'Carrizo' citrange (C. Youtsey, unpublished).

Breeding Objectives for Citrus

Despite the difficulties in citrus breeding, several breeding programmes continue to endeavour to solve important problems limiting citrus production worldwide. Most programmes are separated into scion and rootstock breeding because the objectives of each are usually different, although overall breeding objectives may combine aspects of both programmes to solve more universal problems, e.g. improved freeze-hardiness or virus resistance. Further, most programmes have broad goals as well as specific local priorities. For example, scion breeders emphasize such programmes as improvement of fruit quality (higher colour, TSS, or seedlessness in particular), or selection of early- or late-maturing cultivars to expand the marketing season. Improvement of fruit structure and storeability are also of importance to expand distribution potential. Certainly, freeze-hardiness is of major concern in some subtropical regions such as the southern United States with efforts being made to either increase inherent hardiness, or to change the basic structure of the tree to avoid freeze-damage, i.e. develop a deciduous citrus tree with edible fruit. It is also essential that the scion and rootstock be graft compatible.

Rootstock breeding naturally emphasizes soil-related objectives. However, breeding for improved freeze-hardiness or tree size control are also important objectives in many programmes. Emphasis is being placed on salinity and drought tolerance and resistance to soil-borne organisms such as phytophthora root and foot rot, and citrus and burrowing nematodes. Tolerance or resistance to viruses like the CTV, which is usually a bud-union related problem, is also important. For example, sour orange, which once was the most widely planted rootstock worldwide, is sensitive to CTV and the development of a rootstock with all of the favourable characteristics of sour orange (high quality fruit, freeze-hardiness, adaptability to saline soils) without CTV problems would be highly desirable. Of course, improvements in disease or soil-related characteristics cannot be made at the expense of yield, fruit quality or other desirable characteristics. Moreover, rootstocks should have seedy fruit with a high degree of nucellar embryony to facilitate standard clonal propagation by seed on a commercial basis (Castle, 1987). Rootstocks may also be propagated clonally by tissue culture when seed is unavailable, although this method is more costly than propagation by seed.

Regional breeding objectives encompass the above, but also consider local climatic and edaphic conditions. In many areas, the most widely grown cultivars are local selections chosen for their adaptability over a long period of time. An excellent example of localized breeding and selection has occurred with the satsuma mandarin (*C. unshiu*) in Japan and China. Over hundreds of years, cultivars have been developed via breeding and selection that differ primarily in their time of maturity often by as little as a few weeks. This procedure has provided a continuous supply of fruit for local and export markets.

Breeding Techniques

Traditional techniques

Most citrus breeding programmes have relied on traditional methods for developing new cultivars or rootstocks based on making controlled crosses by hand and selecting superior types from literally thousands of seedlings in the field (Hearn, 1985). Scion breeding involves selection of parental types with favourable heritable characteristics. Often seed parents are selected that produce only zygotic progeny. Parents are usually selected for high combining ability and thus are more likely to transfer favourable characteristics to their progeny. For example, cultivars such as Murcott and Page orange, while having several favourable traits, have poor combining ability and are thus of low value in a breeding programme (C. J. Hearn, unpublished). After hybrid seeds have matured they are extracted from the fruit and planted in the field. Seedlings are maintained until they begin to fruit which often requires ten years or more. During this juvenile period trees are evaluated for disease and pest resistance, freeze-hardiness (where necessary), and for overall growth characteristics. In general, the population of seedlings is extremely variable in vegetative and fruiting characteristics due to the great heterozygosity of citrus. The next step is to bud seedlings with the desired fruit and tree characteristics onto several rootstocks and transfer them to a field nursery for further evaluation. These trees usually fruit within four to five years, after which further fruit and tree evaluations are made. Finally, budwood of the new cultivars is released to nurserymen for further field testing. This process obviously is extremely time consuming. For example, 'Robinson' tangerine was released by the USDA in 1959. The original hybrid between 'Clementine' mandarin and 'Orlando' tangelo was made in 1942!

Traditional rootstock breeding follows the same general procedure as scion breeding with some exceptions (Barrett, 1985; Hutchison, 1985). Large populations of the initial hybrids are planted-out and observed until fruiting after 5–6 years. Fruit must produce sufficient seed numbers to be usable in the nursery and equally important must have a high degree of nucellar embryony and seedling uniformity. Potential rootstocks with these favourable qualities are then planted-out and screened for resistance to soil-borne problems such as phytophthora and nematodes. In many cases, screening tests are used at an early stage of development to determine nematode and phytophthora resistance. Seedlings that pass these requirements are then budded with a common scion and plants are grown out for an additional 5–10 years to determine if they are compatible with the scion and to monitor rootstock effects on fruit quality, freeze-hardiness (where necessary), yields and disease and pest resistance. This entire process may take over 20 years. Therefore, it is not surprising that citrus breeders have been searching for new methods of shortening this costly and time-consuming process.

Biotechnology

Some recent technological advances may help in partially solving or eliminating some of the limitations involved with classical citrus breeding. Isozyme analysis is useful and reliable for distinguishing hybrids from nucellar seedlings. This method is quite accurate for identifying hybrids, including most intraspecific hybrids, but cannot distinguish between closely related cultivars. Hybrids can be further separated into groups using RFLP analysis (see Taxonomy, p. 18). The RFLP method is used to develop maps of the citrus genome that may be useful in locating genes that have a specific function(s). Marker genes are identified on specific chromosomes, some of which are tightly linked to other genes responsible for factors like disease resistance or freeze-hardiness. By mapping these markers, suppositions can be made as to the inheritance of a particular linked gene. Therefore, the ultimate goal is to identify gene locations using molecular biology techniques and then use this information to identify hybrids with the desirable genes and to discard those that do not. Such a system partially eliminates the need for making numerous crosses and screening many seedlings in order to identify favourable characteristics.

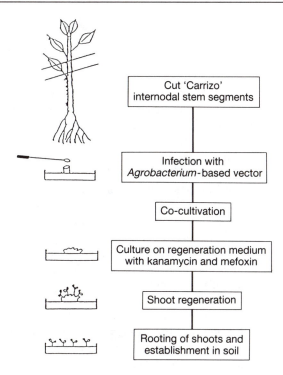

Fig. 2.5. Basic processes involved with genetic transformation of citrus trees. Source: G.A. Moore, University of Florida, Gainesville, USA (unpublished).

This procedure, for example, is being attempted as a means of introducing CTV resistance into sour orange rootstock (G.A. Moore, unpublished). Although the technique, termed 'genetic transformation', is quite complex, its basic features are described in Fig. 2.5. The gene to be introduced is transferred to a bacterial plasmid which serves as a carrier. A plasmid is a circular piece of DNA that can be cleaved at particular sites after which the 'new' gene is spliced into the structure. A specific bacterial strain is then used to introduce the gene and a small amount of bacteria DNA into a tissue culture of the plant, in this instance 'Carrizo' citrange. This DNA is incorporated into the 'Carrizo' DNA. The 'Carrizo' tissue then becomes 'transformed' and begins producing specific proteins coded by the introduced DNA, hopefully without producing any unwanted, undesirable changes in the plant. The transformed plant tissue is cultured on a selective media until shoots are regenerated. Shoots are then rooted in media, after which the transformed plant may be transferred to the greenhouse or field. Such transformed plants have been produced in tobacco, petunia and many other crops including citrus. Plants may also be transformed by introducing DNA directly into the target cell using a high velocity 'gene gun'. In this instance, pieces of DNA are literally fired into the target cell thus eliminating the use of a bacterial strain for introduction. This technique has yet to be successfully used with citrus trees.

The above molecular techniques, nevertheless, are technically difficult and expensive. Consequently, other methods have been developed which are currently being used to circumvent problems associated with conventional breeding. For example, some genera closely related to true citrus may have characteristics that would be desirable if introduced into citrus. An example is the phytophthora and citrus nematode resistance found in *Severinia buxifolia*, a genus within the *Citrus* subtribe, but not in the true citrus group. Since *Severinia* and sweet oranges are sexually incompatible, another system had to be devised to produce hybrids. In the 1980s techniques were developed to circumvent the sexual process by using protoplast fusion (Grosser and Gmitter, 1990). Protoplasts (individual cells lacking the cell wall) are usually extracted from ovule tissue from one and leaf tissue from the other of the two parental species. The protoplasts are cultured on artificial media in the laboratory under carefully controlled conditions. Protoplasts of the two species are placed in a mixture containing polyethylene glycol, which, when removed from the media, causes aggregation of the protoplasts. Protoplast fusion occurs allowing for the unrestricted movement of genetic material between sexually incompatible species. The newly fused protoplasts are grown on selective media until embryos develop. The embryos are nurtured into plantlets and then can be transferred to the greenhouse where the plant is allowed to grow before transfer to the field for further testing. Protoplast fusion, like classical field breeding, is also a long-term proposition but offers a viable alternative to the problem of incompatibility in citrus. Hybrids developed in this manner are tetraploids (4×), however, and may not develop in the same manner as their diploid counterparts.

Mutation and selection

Citrus and related genera are, in general, extremely prone to mutate, although the extent of mutability varies among species. Mutations may occur in single buds or limbs, as a portion of an entire tree, or as the entire tree. Most mutations produce unfavourable or undesirable traits such as abnormally thick peels or dried sections within the fruit. Mutations may also arise, however, that produce earlier- or later-maturing fruit, or fruit with more acceptable juice colour or fruit characteristics.

There are also man-made mutations. Budwood or seeds are treated for short periods with gamma irradiation to induce mutations without killing the cells (Hearn, 1984, 1986). In general, seeds (LD_{50} = 0.1–0.15 Gy) are more tolerant of gamma irradiation than buds (LD_{50} = 0.05–0.09 Gy). The resulting seeds or buds are planted out or propagated by tissue culture techniques. A well-known example of the success of seed irradiation resulted from the 1959 irradiation of 'Hudson' red grapefruit seed by R. Hensz in Texas. Although many undesirable plants resulted, 'Star Ruby' grapefruit was selected in 1970 for its extremely deep-fleshed fruit colour. Other examples include the development of seedless oranges and grapefruit from seedy cultivars. The disadvantage, naturally, to this approach is the random chromosome damage can lead to undesirable characteristics and the need for large populations of plants and long-term field selection of the resultant mutations. Additionally, such mutations may also be unstable.

Chimeras

A chimera is a mixture of tissues of genetically different constitutions on the same part of the plant. This difference usually arises due to mutation, but also may occur due to irregular mitosis, somatic crossing-over of chromosomes, or artificial fusion of unlike tissue resulting from grafting (man-made). Several types of chimera exist in plants which vary by virtue of the relative positions of the dissimilar tissues within the plant organ. The most commonly found forms in citrus are sectorial, periclinal and mericlinal chimeras. Sectorial chimeras involve dissimilarities among one or more sections within the tissue. They are most apparent in fruit where a specific region not always corresponding to a segment is of dissimilar tissue. Commonly, these sections will have different coloration or degrees of resistance to disease or mite damage. A sectorial chimera usually extends from the peel into the seed; consequently, seeds selected from these sectors are genetically similar to other tissues of the chimera. Chimeral selections from citrus with this characteristic have been made and are being further evaluated. Periclinal chimeras consist of an outer sector surrounding an inner sector of dissimilar tissue. Again, using fruit as an example, the peel tissue (exocarp) is of one type and the other internal tissues of other dissimilar types. In this instance, a plant grown from seed is genotypically like the inner, not the outer, tissue. Researchers are exploring tissue culture methods of removing cells from the outer tissues and culturing them *in vitro* with the idea of producing embryos and eventually plants having the genetic characteristics of the outer tissue. This method has been successful using juice vesicles, but has yet to be perfected for peel tissue. Such a technique is potentially valuable in selecting and isolating tissues with superior colour or resistance to diseases, mites or insects. Mericlinal chimeras consist of incomplete sectoring not affecting the internal sectors of the tissue. Thus, although external tissues show differing characteristics, e.g. darker peel coloration, the seeds will produce plants similar to the inner tissues. For example, seeds of 'Pink Marsh' grapefruit will produce white 'Marsh' trees (mericlinal); whereas, seeds of 'Redblush' grapefruit produce 'Redblush' trees (sectorial). Chimeras are common in grapefruit and are important sources of pink and red peel and flesh coloration (see section on grapefruit, p. 36).

3

ENVIRONMENTAL CONSTRAINTS ON GROWTH, DEVELOPMENT AND PHYSIOLOGY OF CITRUS

Citrus may be grown successfully over a moderately wide range of environmental and edaphic conditions. Most commercial production, however, is limited to regions between 40° north–south latitudes where minimum temperatures are generally greater than −7°C. Several microclimatic regions exist within these latitudes and all of the citrus production occurs only in tropical or subtropical climatic regions. For example, many areas in the eastern United States south of 40°N latitude are too cold for citrus production. Environmental factors associated with these climates have a pronounced influence on growth, development and yields of citrus trees, and are largely responsible for the range of yields in mature orchards that is as high as 100 tonnes ha^{-1} in the subtropics to as low as 15 tonnes ha^{-1} in the tropics.

TROPICAL REGIONS

Tropical regions lie between 23.5° north and south of the equator where average annual temperature is above 18°C. Minimum temperatures never fall below 0°C except at the highest elevations. Tropical regions, especially those within 10° north or south of the equator, experience small fluctuations in daylength at all elevations and diurnal temperatures in low and mid elevations. Local climatic conditions vary considerably within the tropical regions based on elevation or proximity to water or mountain ranges which affect wind patterns and rainfall. Therefore, tropical regions may be further subdivided into lowland, midland or highland tropics and into wet (humid) or dry (arid or semiarid) regions.

Lowland tropical regions are located between sea level and about 500 m elevation and have the highest average temperatures and thus the greatest annual heat unit (hu) accumulation. The elevation separating the three regions is not clearly defined and may vary related to average temperature in each zone. An average temperature of less than 24°C, for example, is

Table 3.1. Data on temperature and annual heat units (above 12.5°C) for various citrus-growing areas.

Area and location	Latitude	Elevation above sea level (m)	Annual heat units (°C)	No. of months with average temp.	
				<12.5°C	<17.5°C
Tropical regions					
Trinidad (Piarco Airport)	10°40N	10	5000	0	0
Colombia (Arcataca)	10°30N	30	5500	0	0
Colombia (Giradot)	4°20N	400	5700	0	0
Colombia (Palmira)	3°30N	1000	3500	0	0
Colombia (LaFlorida)	4°40N	1800	1700	0	10
Ecuador (Santa Ross)	3°30S	10	4400	0	0
Ecuador (Conocoto)	0°15S	2200	1000	0	11
Kenya (Mombasa)	4°00S	20	5200	0	0
Kenya (Nairobi)	1°20S	1600	2500	0	1
Uganda (Jinja)	0°30N	1100	3330	0	0
Sri Lanka (Mannar)	9°00N	30	5700	0	0
Sri Lanka (Nuwara Elyia)	7°00N	1900	1000	0	0
Subtropical regions					
Spain (Valencia)	39°30N	30	1600	3	6
California (Riverside)	34°00N	260	1700	3	6
California (Indio)	33°40N	−10	3900	1	4
Israel (Degania)	32°40N	−200	3600	0	4
Israel (Rehovot)	31°50N	50	2600	1	4
Florida (Orlando)	28°40N	30	3700	0	2
Texas (Weslaco)	26°05N	40	3900	0	2
Brazil (Limeira)	22°30S	700	3000	0	1

Source: Mendel (1969).

necessary to induce flowering and may also be used to separate lowland from mid- and highland regions. Heat units are calculated as the amount of time (h) multiplied by the average temperature difference from the minimum for citrus vegetative growth of 12.5°C (Mendel, 1969). For example, a growing region with an average temperature of 18.5°C would accumulate 180 hu for a 30-d month (18.5 − 12.5 = 6.0, 6.0 × 30 = 180 hu). Heat unit accumulation is strongly correlated with growth rate and fruit quality provided that water and nutrients are not limiting. Annual heat unit accumulation in lowland tropical regions like Mombasa, Kenya, Giradot, Colombia, or Mannar, Sri Lanka, is greater than 5000 (Table 3.1). Such high heat unit accumulation increases respiration which decreases fruit solids and acid levels. In moving to mid-elevation tropical areas for citrus (500–1500 m) average temperatures and annual heat unit accumulation are less. For example, annual heat unit accumulation at Palmira, Colombia (elevation 1000 m), is 3500 hu, whereas in highland areas (1500–2500 m) such as Conocoto, Ecuador, annual heat unit accumulation is only 1000 hu. Average annual temperatures at elevations >1500 m often are <12–13°, which is at or below minimum levels for tree growth. Citrus trees are seldom grown at elevations above 2600 m and fruit are used for local consumption only (Camacho-B., 1981). Average annual temperatures may be <10°C under these conditions and are not conducive to tree growth or development of high quality fruit.

Low, mid and high tropical regions vary not only in temperature (heat unit accumulation) but also in rainfall and interception of sunlight. Many lowland areas such as those in parts of Costa Rica, Ecuador and Malaysia are characterized by high relative humidity (RH) and rainfall. However, it is also common to find distinct wet–dry cycles in equatorial Africa, Central America and particularly on islands such as Jamaica. In many tropical regions two distinct wet–dry cycles occur. With the onset of the rainy season the greatest number of flowers are produced, although some flowering occurs throughout the year. Mid-elevation tropical regions also vary in rainfall and humidity with distinct wet–dry seasons such as those in areas of Central America, Venezuela and Colombia. Highland and some midland areas are often distinctly different from the other regions due to the regular presence of ground fog. Fog reduces light intensity during the day, thus reducing temperature and net CO_2 assimilation. At night fog also may reduce longwave radiation losses from the soil, thus increasing temperatures. Moreover, although light intensity may be reduced due to the fog, ultraviolet light may be very intense due to reduced particulate matter in the atmosphere at these high elevations. Ultraviolet light at high levels may cause leaf distortion and reduce growth of citrus trees at these high elevations (R.H. Biggs, unpublished).

SUBTROPICAL REGIONS

The heat unit concept also applies when comparing tropical regions with subtropical ones. Subtropical regions are located between 23.5 and 40° north and south latitude (although not all areas within these latitudes are suitable for growing citrus) and encompass the major citrus-growing regions in the world, viz., portions of Brazil, the United States, Spain, Italy, Japan, Israel, Argentina and much of Mexico and China. Cassin et al. (1968) further divides the subtropics into those areas between 30 and 40° latitude and those between 23.5 and 30° latitude (semi-tropics). Annual heat unit accumulation also varies considerably within the subtropics from 1600 hu in Valencia, Spain, to 3900 hu in Indio, California, and Weslaco, Texas, USA (Table 3.1). Heat unit accumulation in intermediate subtropical and midelevation tropical regions such as Orlando, Florida (3700), and Palmira, Colombia (3500), is intermediate to low tropical and Mediterranean subtropical areas. The distribution of these heat units is more uniform at the intermediate locations allowing for greater net CO_2 assimilation during the winter. Moreover, subtropical regions tend to have more months of average temperatures less than 17.5°C than tropical regions, except at high elevations (Table 3.1).

Subtropical regions are characterized by mean annual temperatures between 15 and 18°C but have greater diurnal temperature fluctuations than tropical regions. Moreover, many subtropical areas are exposed to temperatures below 0°C on a regular basis with temperatures as low as −7°C. Minimum air temperatures of −10°C have occurred in Florida and central China on rare occasions. These low temperatures have had a devastating effect on citrus production in the United States over the years but particularly during the 1980s and also have periodically damaged fruit and trees in parts of Europe, northeastern Mexico, Australia, Argentina and central China. In contrast, freeze-damage rarely occurs in the most productive growing regions of Brazil (São Paulo state). Although freezing temperatures occur in Japan and central areas of China, rarely do extensive crop or tree losses occur due to regional microclimatic effects such as proximity to large bodies of water that moderate temperatures. Also, extended periods of cool temperatures occur in these locations prior to freezes allowing maximum freeze-hardiness development. Japan's citrus industry is located between 35 and 40°N latitude, yet the citrus-growing region of Texas (USA) at latitude 25°N has experienced far more devastating freezes over the years. High-pressure arctic air masses develop in Canada and move unimpeded across the central plains of the USA and into Texas and Florida causing the advective freezes that have been devastating to much of their citrus industries. Furthermore, most of Japan's citrus industry consists of satsuma mandarin on trifoliate orange rootstock, which when fully freeze-acclimated, is an extremely freeze-hardy combination.

Subtropical regions may be further subdivided into humid, semiarid and arid regions. The humid regions such as Brazil and Florida have relatively high humidity and rainfall and smaller fluctuations in diurnal temperatures than in semiarid and arid regions such as southern California (USA), Spain, Italy, Greece, Israel, South Africa and Australia (New South Wales, Murray River). These latter areas, among others, have what is termed a 'Mediterranean-type' climate, which is characterized by hot days, cool nights and low RH and winter rainfall. Although humid subtropical regions have greater annual rainfall than Mediterranean-type climates, rainfall distribution may be seasonal, thus producing drought conditions at certain times. Citrus trees in semiarid or arid regions such as those in southern California or southeastern South Africa are exposed to intense, dry winds which may reduce tree vigour and more importantly cause wind scarring of the fruit. In South Africa, wind scarring is the major cause of peel blemishes which prevents fruit from being marketed fresh. Wind scar is also the major fruit blemish in humid subtropical areas like Florida.

It is logical to expect that such pronounced differences in heat unit accumulation, humidity, rainfall and wind will have a marked effect on citrus tree growth, development, productivity and fruit quality. These environmental factors also influence incidence of pest and disease problems and thus indirectly affect cultural practices within a region.

ENVIRONMENTAL CONSTRAINTS ON VEGETATIVE GROWTH AND DEVELOPMENT

Seed Growth and Development

The citrus seed (Fig. 3.1a) consists of a seed coat surrounding a much reduced nucellus and endosperm (exalbuminous). The seed contains two cotyledons and from one to as many as seven embryos. Only one embryo is derived from sexual fusion of the sperm and egg cells with additional embryos originating from nucellar tissue which is genetically the same as the diploid maternal tissue. Nucellar embryony is very rare in plants and is of importance in production of true-to-type, virus-free rootstocks (Chapter 4).

Seed germination commences with the emergence of the radicle (primary root) through the micropylar (pointed) end of the seed and is dependent on moisture and temperature. Seed germination is hypogeal, i.e. the cotyledons remain underground. The temperature for first emergence of the radicle ranges from 9 to 38°C and varies with cultivar (Fig. 3.2) (W.J. Wiltbank and N. Khoi, unpublished). For example, seeds of Poncirus trifoliata will emerge at 9°C, while those of rough lemon require a minimum of 15°C. This difference is probably related to the tropical origin of rough lemon (India) as compared with the subtropical origin of P. trifoliata (northern China). Days to first emergence range from nearly 80 at 15–20°C to fewer than 14–30 d at the

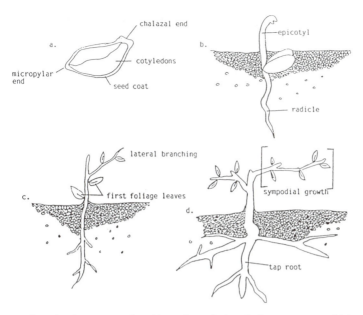

Fig. 3.1. Life cycle of a citrus seedling. (a) Seed morphology before emergence. (b) Seedling growth about two weeks after radicle and epicotyl emergence. (c) Development of lateral branches. (d) Development of a sympodial growth habit.

optimum range of 30–35°C for most cultivars. Interestingly, removal of the seed coat prior to emergence shifts the curve so that emergence occurs at lower temperatures and more rapidly than when seed coats are present. Seed coat removal also improves percentage emergence. This practice is being used by some citrus nurserymen to improve percentage emergence and hasten emergence time, but is only practical where labour costs are low.

Light intensity does not affect emergence or germination, but seedlings developing in the dark will be etiolated and spindly.

The Juvenile Plant

Citrus seedlings undergo a relatively long juvenility period (time until first flowering) following germination, depending on species and growing conditions. The seedling typically grows from a single apical meristem during the first few weeks after germination (Fig. 3.1b) but then begins to produce lateral meristems in the leaf axils from which subsequent branching may occur. Juvenile citrus trees are thorny (modified leaves) and generally have a very upright, unbranched growth habit. As the tree grows the apical meristem abscises and lateral buds break along the central axis. New shoots emerge

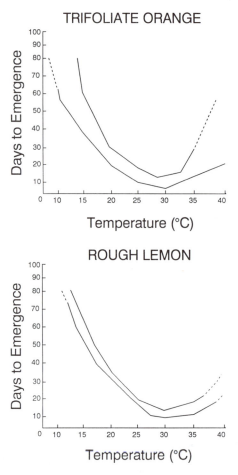

Fig. 3.2. Effect of temperature on seedling emergence of two citrus species. For each graph: upper line = + seed coat and lower line = − seed coat. Source: W.J. Wiltbank and N. Khoi, University of Florida, Gainesville, USA (unpublished).

from vegetative lateral buds (Fig. 3.1c) and in turn the apical meristem of these shoots will abscise and other lateral buds emerge. This zig-zag growth pattern, termed sympodial growth, is characteristic of most citrus and related species (Fig. 3.1d). Since the terminal bud abscises, citrus trees have a determinate growth habit.

As the new shoots emerge the apical meristem produces leaf primordia at regular intervals around the stem as it elongates. Citrus species have characteristic patterns of phyllotaxis. For example, sweet orange has a 3/8 phyllotaxis and grapefruit a 2/5 phyllotaxis. In 3/8 phyllotaxis, eight leaves are produced around the stem in a spiral pattern with the eighth leaf emerging directly above the first leaf of the spiral. This occurs within three revolutions around the stem. For grapefruit, five leaves are produced within two revolutions around the stem. These patterns are then repeated as the shoot develops. Although phyllotaxis occurs in a somewhat regular manner within a citrus species, environmental factors may alter the pattern, e.g. where shoots have developed under water stress or excessively vigorous growing conditions.

Duration of juvenility varies within species and with environmental factors. Generally, the juvenility period is inversely related to tree vigour and heat unit accumulation, again provided other factors are not limiting. Vigorous species such as limes and lemons have juvenility periods of less than 2 years under subtropical growing conditions, while juvenility periods of 5 to as long as 13 years may occur for mandarins, sweet oranges and grapefruit when grown from seed. Marcotts (air-layers) of 'Tahiti' lime often produce fruit within 1 year of planting. Duration of juvenility is dramatically affected by temperature, moisture, and in some cases edaphic and cultural conditions. For example, in lowland tropical areas with high rainfall the juvenility period is considerably shorter than in arid subtropical regions with suboptimal irrigation.

Citrus breeders in particular are interested in shortening the juvenility period to decrease the time required to breed and select new cultivars. Calamondin and 'Key' lime have been forced to flower in fewer than 18 months from seed by growing them in controlled climate chambers at 30°C days, 25°C nights and 16 h daylengths (Snowball et al., 1988). Under these conditions plants grew continuously (not in distinct growth flushes) attaining a height of 2 m. Growth was then arrested using paclobutrazol (a growth retardant) and extensive lateral branching and flowering occurred. Unfortunately, this technique does not produce flowers for all citrus species but appears to be useful for shortening the juvenility period of some species. An alternative method of shortening the juvenility period is to grow seedlings in the greenhouse as single, long stems. The top of the stem is bent to the base, thus forcing it to flower.

The Budded Plant

The juvenility period is rarely a problem in most citrus-growing regions since most commercial citrus trees in the world are grown as plants consisting of two parts, the scion and the rootstock. The scion originates from budwood of mature bearing trees and is budded onto a rootstock (Chapter 5). This practice usually promotes earlier fruit production than found in seedlings as well as taking advantage of scions and rootstocks with desirable and uniform fruit and tree characteristics (Chapters 2 and 4).

Growth and development of budded nursery trees is also dependent on environmental factors. For example, nursery trees grown in Riverside,

California, develop to a marketable size in about 30 months at 1700 hu, while the same rootstock/scion combination develops in only 15 months at Mannar, Sri Lanka, at 5700 hu (Mendel, 1969). Similarly, a 3-year-old 'Washington' navel orange tree attained a height of 4 m under lowland tropical conditions (5700 hu), while a 4-year-old tree attained only 3 m under highland conditions (2000 hu). The same type of disparity in growth rates occurs when comparing citrus trees growing under arid subtropical and humid subtropical growing conditions. Citrus tree growth rates are much slower in Valencia, Spain (semiarid subtropical), than in Orlando, Florida (humid subtropical), for example. This difference occurs not only due to differences in heat unit accumulation (1600 vs 3700) but also to the greater rainfall and relative humidity in Orlando.

Light intensity and quality also affect growth and development of the vegetative citrus tree in several ways. Light intensity has a direct effect on net CO_2 assimilation and an indirect effect on leaf temperature. Growth of vegetative citrus trees is closely related to net CO_2 assimilation provided that other factors such as temperature, nutrition and water are not limiting. Net CO_2 assimilation increases linearly as photosynthetic photon flux (PPF) increases from 0 to about $700\,\mu mol\,m^{-2}\,s^{-1}$ and plateaus above this level known as the light saturation level (Syvertsen, 1984) (Fig. 3.3). A PPF of $2000-2200\,\mu mol\,m^{-2}\,s^{-1}$ represents full sunlight near sea level; however, cloud cover can significantly reduce PPF to below light saturation in many tropical locations. Therefore, maximum net CO_2 assimilation for most citrus species is achieved at 30–35% of full sunlight. The longer a citrus tree is at or above light saturation, potentially the greater the net CO_2 assimilation, again provided that water, temperature, nutrition or other factors are not limiting photosynthesis. Therefore, vegetative growth of citrus trees is usually greatest where daylengths are moderate to long (>12 h), maximum heat units are attained, and water is not limiting, e.g. in high rainfall, low tropical regions. In contrast, it is not uncommon for PPF to be $100\,\mu mol\,m^{-2}\,s^{-1}$ or less within the canopy of a mature citrus tree in arid regions of South Africa. Consequently, net CO_2 assimilation levels are often less than $2\,\mu mol\,m^{-2}\,s^{-1}$ with a concomitant reduction in flower number and yield in this region.

In many arid regions such as Israel or Indio, California, extremely high light intensity may decrease net CO_2 assimilation due to increased radiation load on the leaf. Under extreme conditions leaf temperature may be 7–10°C greater than air temperature and may approach 55°C. Optimum temperatures for net CO_2 assimilation for citrus are species-dependent and range from 28 to 30°C in humid air (Kriedemann and Barrs, 1981) with temperatures greater than 35°C severely limiting the activity of ribulose 1,5-bisphosphate carboxylase/oxygenase (RuBisCo), and possibly causing midday stomatal closure. At low humidity, temperature optima range from 15 to 22°C. As temperatures increase leaf-to-air vapour pressure deficit (VPD) also increases, lowering stomatal conductance. Citrus trees have moderate to

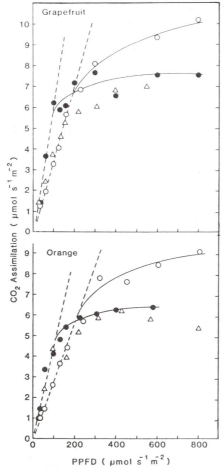

Fig. 3.3. Light responses of net CO_2 assimilation rate and apparent quantum yield (μmol CO_2 assimilated per incident μmol PPF, dashed lines) of grapefruit and orange leaves grown for 5 months under low (●) or (○) PPF and after being moved from low into high (△) PPF for 14 d. Curved lines were fitted by eye. Source: Syvertsen (1984).

low net CO_2 assimilation compared with other fruit trees species, such as cherry or apple; values range from 9 to 12 μmol m^{-2} s^{-1} of CO_2 assimilated under optimum environmental conditions (Syvertsen, 1984) compared with 25–30 μmol m^{-2} s^{-1} for cherry. Lower than optimum temperatures also decrease net CO_2 assimilation probably due to effects on VPD and enzyme activity. Net CO_2 assimilation, however, did not exceed 6 μmol m^{-2} s^{-1} under the high VPD conditions found in arid climates (such as northeastern South Africa). Nevertheless, annually citrus fixes a moderately high amount of CO_2 because it is an evergreen, unlike cherry or apple which are deciduous.

Freezing, but not lethal low temperatures, also significantly decrease net CO_2 assimilation for several days following a freeze by causing stomatal closure and decreasing activity of enzymes involved in photosynthesis (Young, 1969).

There is also some suggestion that light quality affects vegetative growth of citrus trees. Studies from Israel (Mendel, 1969) suggest that shoot elongation is enhanced by far-red light and inhibited by red light. Generally, far-red light levels are greatest inside the tree canopy. High ultraviolet-B light has an inhibitory effect on shoot growth of some citrus species (R.H. Biggs, unpublished).

Shoot and leaf growth

Shoot elongation of citrus trees usually occurs in two to five distinct growth flushes annually in subtropical regions but may occur on a continuous basis in lowland tropical areas and in some coastal subtropical areas, particularly for lemons and limes (Mendel, 1969). Commencement of shoot growth is regulated by temperature (>12.5°C) in subtropical regions and by availability of water in tropical regions (Cassin *et al.*, 1968). Seasonal cumulative shoot elongation or dry matter accumulation is generally greatest under moderate, consistently long days and high mean day and night temperatures typical of low tropical areas.

The distribution and extent of shoot growth is also affected by temperature. In subtropical regions, the spring growth flush, which generally occurs in March–April in the northern hemisphere and September–October in the southern hemisphere, usually occurs from many growing points producing many shoots with short internodes. Mean temperatures generally range from 12 to 20°C during this period. In contrast, the summer growth flush (which occurs in June–July in the northern hemisphere and January–February in the southern hemisphere) occurs from fewer growing points but produces shoots with longer internodes. Temperatures during this period range from 25 to 30°C. Late summer or autumn growth flushes also tend to occur from few growing points.

Potentially shoots are produced throughout the season in tropical regions due to high mean temperatures year-round, provided that water is not limiting as occurs in tropical regions with distinct wet–dry climatic cycles. Shoots are generally produced from many growing points with internode length dictated by plant water status, as well as temperature. Therefore, a citrus tree appears to have a genetically predetermined level of growth that can occur in a few major flushes or almost continuously as minor growth flushes.

Citrus trees, being evergreen, are densely foliated within a few years in the field which causes extensive shading of interior portions of the canopy. For example, leaf numbers and surface areas of a citrus tree growing in California increased from 16,000 and 34 m^2 at age 3 years, to 37,000 and 59 m^2 at 6 years, 93,000 and 146 m^2 at 9 years and 173,000 and 203 m^2 at 29 years (Turrell, 1961). The rate of canopy development may differ from these values depending on climate. Many mature trees produce over 350 m^2 of leaf surface area with a leaf area index of 12. Leaf area index (LAI) is a measure of the total leaf surface area per unit of land covered by the plant. In large trees most of the fruiting occurs in the outer metre of the canopy because radiant energy is reduced to nearly zero at depths into the canopy greater than this and PPF is typically less than 100 μmol m^{-2} s^{-1}. LAI is curvilinearly or log-linearly related to photosynthetic photon flux in the canopy (Jahn, 1979). Obviously, at high LAI, light severely limits shoot and flower production (PPF <50 μmol m^{-2} s^{-1}) in the interior of the canopy, and most new growth occurs in the outer canopy. The problem of shading becomes even more acute for high-density plantings. Therefore, a judicious yearly pruning or hedging and topping programme is essential to maintain light penetration and promote fruit production in mature trees (see Chapter 5).

Newly developing leaves are generally carbon importers until fully expanded about 4–6 weeks after full bloom (Erner and Bravdo, 1983). Stomata, which occur on the abaxial surface, are not fully developed and stomatal control over transpiration is poor. Net CO_2 assimilation continues to increase until about 6 months later at which time net CO_2 assimilation becomes stable until the latter stage of leaf aging. Healthy citrus leaves can remain on the tree for as long as 3 years depending on tree vigour, but disease, pest pressures and low light levels significantly reduce leaf longevity (Chapter 6).

Root growth

Temperature also regulates root growth and development and water and nutrient uptake. Root and shoot growth have different temperature thresholds with root growth occurring at temperatures higher than about 7°C. Root growth, like shoot growth, occurs in flushes which often but not always alternate with shoot growth flushes (Bevington and Castle, 1985). Studies from Florida suggest that major root growth periods occur from May to June and particularly from August to September in the northern hemisphere (Bevington and Castle, 1985) (Fig. 3.4). Moreover, mean elongation rate for citrus roots is strongly temperature dependent for both pioneer and fibrous roots, showing a linear positive increase in growth from 17 to 30°C.

Water and nutrient uptake rates are also positively correlated with root temperatures. Root hydraulic conductivity, a measure of water uptake, increases considerably from 10 to 30°C (Wilcox and Davies, 1981). Since uptake of nutrients is related to water uptake and respiration, it is also temperature dependent. Winter chlorosis of citrus leaves occurs in some subtropical regions possibly due to a soil temperature-induced reduction in nutrient uptake.

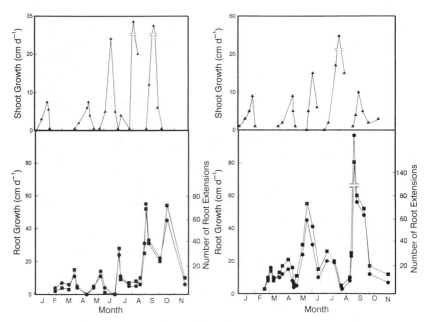

Fig. 3.4. Typical pattern of root and shoot growth during 1982 for 'Valencia' orange trees on rough lemon (left) and 'Carrizo' citrange (right) rootstocks. Data are shown for replicates of each rootstock. Symbols: ▲, shoot growth; ■, root growth; ●, number of roots. Source: Bevington and Castle (1985).

Plant and soil moisture status also affects root growth. Soil matrix potentials less than −0.05 MPa inhibited root elongation and the production of new roots (Bevington and Castle, 1985). Upon rewatering elongation rate increased but the primary increase in root area was due to production of new roots. Root volume of young citrus trees and production of new roots decreased significantly as soil moisture percentage decreased to less than 45% of available water for a sandy soil in Florida (Marler and Davies, 1990). Sufficient soil moisture is also necessary to support functions of non-growing roots. Nevertheless, excessive water for even as little as a few days may cause root death at high soil temperatures due to production of sulfur dioxide by soil-borne bacteria (see Flooding, Chapter 5).

ENVIRONMENTAL CONSTRAINTS ON FLOWERING AND FRUITING

Environmental factors, particularly water and temperature, regulate the time and extent of flowering in citrus trees. Therefore, intensity and duration of flower production also varies with climatic region. Moreover, environmental factors regulate the type of flowers produced, their distribution on the tree, the percentage of fruit set and ultimately the resulting yield. Flowering of citrus consists of induction and differentiation (evocation) periods which precede anthesis.

Flower Induction and Differentiation

Flower bud induction commences with a cessation of vegetative growth during the winter 'rest' (nonapparent growth) period in subtropics or dry periods in tropical regions. Generally on mature trees, shoot growth ceases and root growth rate decreases as temperatures decrease into the winter even though temperatures are not below 12.5°C. During this period vegetative buds develop the capacity to flower. Therefore, induction involves the events directing the transition from vegetative growth to production of infloresences (Davenport, 1990). Davenport (1990) and Garcia-Luis et al. (1992) have proposed that bud initiation may precede induction, but experimental evidence for this is limited. Cold and water stress are the primary inductive factors, with cold being the primary factor in subtropical climates and water stress in tropical climates. Temperatures below 25°C for several weeks appear to be required for induction of flower buds in significant quantities (Inoue, 1990). In the field drought periods longer than 30 d are required usually to induce a significant number of flower buds. The degree of induction is proportional to the severity and duration of stress (Southwick and Davenport, 1986). Water stress has been used as a practical means of inducing flowering in citrus for many years. In Italy, water is withheld from lemon trees during the summer until trees become severely water stressed. During this time flower buds are induced but rarely develop. The trees are then irrigated 'forcing' (forzatura) them to flower in the fall thus producing a crop the next summer ('Verdelli' lemons). Generally, trees will flower 3–4 weeks after irrigating. This method has also been used to force off-season flowering in Israel and Spain and for 'Tahiti' limes in Florida. Application of gibberellic acid during the flower induction period will prevent induction and inhibit subsequent flowering (Monselise and Halevy, 1964; Davenport, 1990).

Differentiation involves the histological and morphological changes in the vegetative meristem to become a floral meristem (Davenport, 1990). The dome-shaped apical meristem broadens and flattens and organogenesis begins with formation of the sepal primordia followed by carpel development. Once sepal primordia have formed the flower bud will not revert to the vegetative condition even with application of gibberellic acid (Lord and Eckard, 1987). The anatomical status of the terminal apex determines the sequence and disposition of the lateral buds (Lord and Eckard, 1987). If the terminal apex forms sepals, the lateral buds will also form flowers, if the apex forms leaves the lateral buds will form thorns, which in this instance are modified axillary

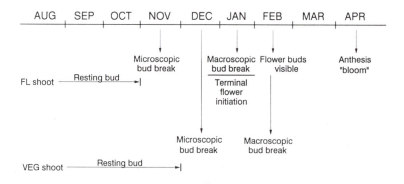

Fig. 3.5. Diagrammatic representation of the progression of events leading to anthesis in citrus in a subtropical climate in the northern hemisphere. Source: Lord and Eckard (1985).

flowers, not leaves (Lord and Eckard, 1987). The rate of flower development from bud break to anthesis is independent of flower position or inflorescence type and is positively correlated with degree days (Lovatt *et al.*, 1984). The typical progression of flower induction through anthesis in subtropical California (northern hemisphere) is presented in Fig. 3.5 (Lord and Eckard, 1985). The resting bud undergoes microscopic bud break (bud scale loosening) as early as November and December; macroscopic bud differentiation occurs during December and January, with flower buds becoming visible as early as February. Anthesis then occurs in April. However, time of flower bud induction and anthesis varies considerably from season to season depending on temperature and water availability. In subtropical regions such as Florida, cold induction may also start as early as November, but as in 1993, induction can occur as late as March with flowering occurring after April 1.

Anthesis

Anthesis (flowering) occurs after induction and differentiation when favourable temperature and soil moisture conditions exist. Minimum threshold temperature for flowering is 9.4°C, or considerably lower than the minimum for vegetative growth (Lovatt *et al.*, 1984). Citrus flowers are borne on cymes with the terminal flower bud breaking first followed by the most basal flower bud on the shoot. The second flower bud to the apical (terminal) position is the last to open, probably due to apical dominance (Jahn, 1973). Lord and Eckert (1985) also observed that the terminal flower was first to open for

Table 3.2. Number of fruit set by position on the inflorescence and by the number of flowers on the inflorescence of 'Hamlin' and 'Valencia' oranges, 1970.

Fruit position	No. of flowers/inflorescence									No. of flowers	Per cent fruit set
	1	2	3	4	5	6	7	8	9		
'Hamlin'											
1[1]	40	1	1	1						731	5.9
2		14	18	4	10	5	3	1	1	327	17.1
3			3	3	3	2	2	2		257	4.7
4				6	3	3				185	7.6
5					2	1				127	2.4
6						1				56	1.8
7										21	0.0
8										8	0.0
9										1	0.0
No. of flowers	404	140	216	232	355	210	91	56	9	1713	
Per cent fruit set	9.9	10.7	9.7	6.0	4.8	5.7	4.4	3.6	11.1		7.5
'Valencia'											
1[1]	4	2		2		1				640	1.0
2		9	6	7	7	3				337	9.5
3			4	4	3	1	1			268	4.9
4				3	6	3		2		212	6.6
5						3				130	2.3
6						2				68	2.9
7							1			19	5.3
8										4	0.0
No. of flowers	303	138	168	328	310	294	105	32		1678	
Per cent fruit set	1.3	8.0	6.0	4.9	5.2	4.4	1.9	6.3			4.4

Source: Jahn (1973).
[1]Position nearest the apex of the inflorescence.

navel oranges. Lateral flower buds in positions 6 and 7 opened next and inhibited bud break of more basal buds. Flower size generally decreases from the terminal to the last flower to open. Thus, the second flower position below the apex usually produces the smallest flower but also has the highest per cent fruit set on the shoot (Jahn, 1973) (Table 3.2). Late-opening flowers grow faster and persist longer than early-opening flowers (Lovatt *et al.*, 1984).

Five basic types of growth arise during flowering: (i) generative shoots (leafless or bouquet bloom) having flowers only borne on previous season's growth; (ii) mixed shoots having a few flowers and leaves; (iii) mixed shoots having several flowers and a few large leaves; (iv) mixed shoots having a few flowers and many leaves; and (v) vegetative shoots having leaves only. All of the mixed shoots produce flowers and leaves in the new growth flush (leafy blooms). Leafy blooms set a higher percentage of flowers than leafless blooms. Generally shoots with a high leaf:flower ratio, such as category (iv), produce and hold the greatest percentage of fruit to maturity. Vegetative shoots, category (v), produce the greatest shoot growth during the season and the generative shoots the least, with other categories intermediate and related again to the amount of fruit set and carried. It is unclear whether the greater fruit set of leafy blooms is due to increased net CO_2 assimilation and carbohydrate levels provided by the newly developed leaves, or to improved vascular connections to the developing fruit mediated by hormones from the newly developed leaves or a greater sink capacity of mixed buds. Improved vascular connections would reduce water stress in leafy compared with leafless blooms. All these hypotheses have merit, although under some conditions net CO_2 assimilation is similar for both leafy and leafless blooms and carbohydrate levels have not always correlated with fruit set. Alternatively, newly developing leaves 4–6 weeks after anthesis, may decrease fruit abscission rather than increase fruit set (Erner and Bravdo, 1983).

Generally, more leafless blooms are produced on a tree than leafy blooms and a higher percentage of the fruit produced arise from leafless blooms (Erner and Bravdo, 1983). Percentage of leafless blooms of 'Marsh' grapefruit and 'Valencia' orange trees varied from 55–60 to 70–75 in 2 consecutive years in Florida (Jahn, 1973). There has been interest in finding practical methods of producing a higher percentage of leafy blooms, thus increasing fruit set and potentially yields. The relative abundance of leafy and leafless blooms seems to be related to temperature. Seasons with low winter temperatures of long duration lead to development of more leafless blooms and more leafy blooms occur in seasons with high winter temperatures. Low to moderate temperatures during bloom (<20°C) produce a protracted bloom, while temperatures of 25–30°C produce a shorter bloom period. Relatively high winter temperatures (low induction) also leads to later, extended bloom. Application of urea sprays during flower induction may increase the number of leafy blooms (C. J. Lovatt, unpublished).

Factors Associated with Flowering

Numerous studies have been conducted over the years to determine which physiological factors control flowering in citrus. These have been reviewed in detail by Davenport (1990). The most likely control factors are carbohydrates, hormones, temperature, water relations and nutrition. The carbohydrate theory has its basis on the fact that branch or trunk girdling increases flower induction, fruit set and starch levels in the branch. This probably occurs because girdling inhibits phloem transport of carbohydrates to the roots. In contrast, several studies have found no correlation between starch levels in leaves and twigs and flowering of citrus (Oslund and Davenport, 1987; Davenport, 1990). Carbohydrate levels in roots, however, are in some instances associated with flowering in alternate bearing mandarins. Extremely low levels of carbohydrates in the roots due to excessive crop loads have been associated with limited shoot and flower production. This condition is especially severe in 'Murcott' ('Honey') mandarin trees which in some instances produce so much fruit that root carbohydrates become depleted and trees die ('Murcott' collapse). Roots are also a source of carbohydrates, therefore once again the correlation between carbohydrate levels and flowering may not be causal. A critical level for carbohydrates is at least implied even if higher levels of carbohydrates do not elicit greater flowering.

Hormonal control of citrus flowering has also been extensively studied for many years (Davenport, 1990). Some studies involve application of endogenous hormones to citrus shoots followed by an evaluation of the extent of flowering. For example, application of gibberellic acid (GA_3) to citrus shoots before differentiation inhibits flowering (Monselise and Halevy, 1964). Therefore, GA_3 appears to regulate some aspects of flowering. However, studies on changes in endogenous levels of GA_3 indicate no significant relationship between GA_3 and type of shoot (generative or vegetative) produced (Davenport, 1990). It is possible that gibberellins are involved with flowering through an intermediate process, although there are no definitive studies supporting this idea.

Plant nutrition is directly and indirectly associated with flowering of citrus trees. High levels of leaf N in particular for young citrus trees may induce excess vigour and produce a vegetative rather than flowering tree. In contrast, low leaf N levels promote extensive flowering, although fruit set and yields are poor. Severely N-deficient trees produce few flowers. Therefore, maintaining leaf N levels in the optimum range (2.5–2.7%) produces a moderate number of flowers but the greatest fruit set and yields. Nitrogen as ammonia may directly affect flowering via regulation of ammonia and polyamine levels in the bud (Lovatt et al., 1988). Water or low temperature stress increases leaf ammonia levels and flowering. Moreover, a winter application of urea during the stress period to 'Washington' navel orange trees in California (USA) increased leaf and bud ammonia levels and number

Table 3.3. Effect of low temperature stress and foliar application of urea on the leaf NH_3–NH_4^+ content and on the flowering of 5-year-old rooted cuttings of the 'Washington' navel oranges.

Duration of low temperature stress (weeks)	Leaf NH_3–NH_4^+ content[1]		Average no. of flowers per tree	
	Without urea[2]	With urea[3]	Without urea	With urea
0	456a	–	6a	–
4	559b	928 (166%)	117b	227 (194%)
6	583b	1253 (215%)	131b	310 (230%)
8	672a	900 (134%)	347a	437 (126%)

Source: Lovatt et al. (1988).

[1]Ammonia (μg g^{-1} dry weight) determined during first week after transfer to warmer temperature.

[2]Mean separation by Duncan's multiple range test, 5% level.

[3]Low biuret urea applied at the rate of 1.5 g per tree at the end of the low temperature treatment. Figures in parentheses represent percentage of the value recorded in the trees where no urea was added.

of flowers per tree; the number of flowers produced was positively correlated with the duration of low-temperature induction (Table 3.3). Similarly, spray applications of 1% urea 6–8 weeks before bloom increased flowering and yields of 9-year-old 'Shamouti' orange trees (Rabe and van der Walt, 1992).

It is obvious from the previous sections that the physiological basis of flowering in citrus is not clearly understood, although many pieces of the puzzle have been assembled. Flower buds are induced by cool temperatures or drought stress. Buds differentiate when warm temperatures or soil moisture release the stress. Time to flowering is also temperature dependent. Although these processes are probably under hormonal control, only GA_3 has clearly been shown to play a role (inhibitory). Part of the difficulty in understanding the control of flowering in citrus may be because some buds appear very easy to induce to flower while other buds, even on the same branch, appear to require much more stress to be induced.

Pollination and Fruit Set

Most commercially important citrus species do not require cross-pollination to set and produce a crop. The exceptions to this include some mandarin hybrids such as 'Orlando' tangelo and 'Robinson' tangerine. Pollination is essential, however, for seed production, or in stimulating ovary growth in nearly parthenocarpic cultivars such as Hamlin sweet oranges. Parthenocarpy is the capacity to produce fruit without the stimulus of sexual fertilization. Some strongly parthenocarpic cultivars like Marsh grapefruit, however, produce fruit even if the stigma and style are detached prior to pollination.

Temperature has a significant effect on pollination efficiency, either indirectly by affecting bee activity in the orchard (bees are the primary pollinator for citrus), or directly by affecting pollen tube growth rate. Bee activity in an orchard is minimal at temperatures below 12.5°C. Once pollen grains land on the stigma their germination and growth rates through the style are enhanced at high temperatures (25–30°C) and reduced or totally inhibited at low temperatures (<20°C). Pollen tube growth through the stylar canals may take from 2 d to as long as 4 weeks depending on cultivar and temperature (Frost and Soost, 1968).

Initial fruit set

Initial fruit set, subsequent fruit drop and ultimately fruit yields are affected by several environmental and physiological factors. Most commercially important citrus cultivars bloom prolifically producing as many as 100,000–200,000 flowers on a mature tree; however, fewer than 1–2% of these flowers will produce harvestable fruit (Erickson and Brannaman, 1960) (Table 3.4). An initial drop period occurs from flowering until 3–4 weeks postbloom. A second drop period occurs from May to June in the northern hemisphere and November to December in the southern hemisphere. The initial drop period involves the abscission of 'weak' flowers and fruitlets with defective styles or ovaries, or flowers which did not receive sufficient pollination (where applicable) (Erickson and Brannaman, 1960). During initial phases of abscission (6–8 weeks after bloom) most fruit abscise at the zone between the pedicel and the stem. Citrus fruit have two abscission zones, one at the base of the pedicel and the other at the base of the ovary.

Hormones are probably involved with the capacity of fruit to persist during initial fruit set based on circumstantial evidence. Spray application of GA_3 increases initial fruit set for weakly parthenocarpic species like 'Orlando' tangelo, although GA_3 does not improve fruit set of most other citrus cultivars. Similarly, spray application of benzyladenine (BA), a cytokinin, has been shown to improve fruit set of navel oranges in Spain.

Time of anthesis is also linked to percent initial fruit set. Studies from California suggest that flowers opening early in the bloom period have much lower set than those opening later. Low temperatures early in the season may limit bee activity or pollen tube growth as discussed previously. In addition, fruits developing early during bloom on leafless shoots grow more slowly than those developing on leafy shoots later during bloom due to higher temperatures at that time (Lovatt et al., 1984). Extremely high temperatures (>40°C), however, cause excessive fruit drop particularly for cultivars like navel oranges

Table 3.4. Abscission of reproductive structures per tree from 'Washington' navel and 'Valencia' orange, Riverside, California, 1958–59.

Ovary diam. (mm)	'Washington' navel					'Valencia' orange				
			No. of fruit					No. of fruit		
	No. of buds	No. of flowers	With pedicel	Without pedicel	Total no.	No. of buds	No. of pedicel	With pedicel	Without pedicel	Total no.
<1	59,635	674	1,002	12	61,323	20,020	341	893	173	21,427
2	22,790	14,939	13,953	344	52,026	4,099	565	5,468	2,621	12,753
3	10,804	15,295	25,043	541	51,683	1,365	1,031	11,151	1,880	15,427
4	3,114	2,321	13,469	589	19,493	499	219	7,610	2,533	10,861
5	–	6	3,853	460	4,319	–	7	1,956	2,161	4,124
6	–	–	2,399	634	3,033	–	–	702	1,898	2,600
7	–	–	1,250	643	1,893	–	–	137	1,164	1,301
8	–	–	586	450	1,036	–	–	59	582	641
9	–	–	417	663	1,080	–	–	21	699	720
10	–	–	179	462	641	–	–	5	611	616
11	–	–	77	357	434	–	–	2	453	455
12	–	–	37	275	312	–	–	1	392	393
13	–	–	20	194	214	–	–	5	297	302
14	–	–	7	137	144	–	–		237	237
15	–	–	7	118	125	–	–		158	158
16	–	–	3	79	82	–	–		106	106
17	–	–	2	71	73	–	–		83	83
18	–	–	2	55	57	–	–		65	65

19	—	—	2	40	42	—	—	—	34	34
20	—	—	—	32	32	—	—	—	23	23
>21	—	—	—	232	232	—	—	—	61	61
Total	96,343	33,235	62,308	6,388	198,274	25,983	2,163	28,010	16,231	72,387
Average no. of mature fruit/tree					419					708
Total flower buds/tree					198,693					73,095
Distribution	48.5%	16.7%	31.4%	3.2%	crop=0.2%	35.5%	3.0%	38.3%	22.2%	crop=1%

Source: Erickson and Brannaman (1960).

which are inherently more sensitive to stress than most other citrus species (Davies, 1986a).

Physiological drop

Although the initial drop period in citrus apparently is primarily for physiological reasons, the term 'physiological drop' generally has been reserved for the drop wave that occurs in May to June in subtropical regions in the northern hemisphere and in November to December in the southern hemisphere. Physiological drop is also called June (or November) drop and describes the abscission of fruitlets as they approach 0.5–2.0 cm in diameter. Physiological drop is a disorder most probably related to competition among fruitlets for carbohydrates, water, hormones and other metabolites. The problem, however, is greatly accentuated by stress, especially high temperatures or water deficit. Consequently, physiological drop is usually most severe where leaf temperatures may reach 35–40°C and where water stress is a problem, as in arid regions of southern California or South Africa. One hypothesis is that high temperatures and severe water stress cause stomatal closure with a concomitant reduction in net CO_2 assimilation. Fruit abscission then results because the fruit maintain a negative carbon balance. Overhead sprinkling reduces temperature and has been used in some regions to reduce physiological fruit drop. Differences in hormone or carbohydrate levels also may be involved, although data are lacking. Fruit drop occurs from the abscission zone at the base of the fruit leaving the pedicel attached to the tree temporarily.

ENVIRONMENTAL CONSTRAINTS ON FRUIT YIELDS

Because climate affects flowering, fruit set, fruit drop and cumulative fruit number per tree, it is logical to expect that fruit yields will also be affected. In addition, fruit size, another component of yield, is affected by climate, in particular moisture and temperature. Cultural practices, as well as the choice of cultivar and rootstock (Chapters 2 and 4), and nutrition (Chapter 5), also affect yields but will be discussed in subsequent chapters.

Generally, the years to production of a crop is shortest for the low tropical and humid subtropical regions. There is then a period where tree size (canopy bearing volume) and yield increase asymptotically for all three regions. Yields in low tropical regions become maximum 10–15 years after planting, while yields in the humid subtropical regions continue to increase reaching a maximum at 15–20 years. Yields from the Mediterranean climate also increase gradually peaking around 20–25 years. Orchard longevity is usually greater in Mediterranean-type climates than humid subtropical or low tropical climates. Many orchards in Spain and Italy, for example, are over 200

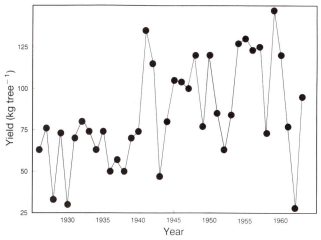

Fig. 3.6. Yields of 'Washington' navel in California over a 38-year period. Source: Jones and Cree (1965).

years old, while most orchards in humid subtropical and low tropical regions are less than 100 years old. Disease and pest pressures account for much of these differences in orchard longevity.

Excluding freeze events, yields of commercially important citrus cultivars vary considerably from year to year, primarily due to climatic factors. Research from South Africa (DuPlessis, 1984) suggests that a large part of the yearly variation in yields of navel oranges is due to climatic factors, in particular temperature during the physiological drop period. Similarly, 38 years' data on navel oranges in California (USA) also clearly demonstrates the irregularly cyclic nature of yields related to climate (Jones and Cree, 1965) (Fig. 3.6). Major factors associated with this variation include temperature during bloom, physiological drop, and throughout the growing season. In general, excessively high temperatures during these times reduced yields. Soil moisture was not a major factor in this instance since most of the orchards were irrigated.

The maximum yield and year-to-year variation obtained from a mature citrus orchard is also a function of climate, although factors such as soil type, cultivar and rootstock selection, technological capacity, and disease constraints also influence yields within a climatic region. Soil moisture as well as temperature become major factors regulating yields in the humid subtropics (Koo, 1963) and tropical areas, and in arid and semiarid regions (Hilgeman, 1977). Furthermore, maximum yields are related to cultivar and rootstock with grapefruit generally being most productive, followed by sweet oranges, lemons and mandarins. Thus, the most productive citrus-growing regions generally occur in the humid subtropics. For example, in Florida average yields for grapefruit and sweet oranges are 41 and 31 tonnes ha^{-1}, respectively. However, maximum yields may exceed 120 tonnes ha^{-1} for grapefruit and 100 tonnes ha^{-1} for sweet oranges. In contrast, per hectare yields in Brazil, also located in the humid subtropics and tropics, is about one-half that of Florida primarily due to a lack of adequate irrigation in most major growing regions of São Paulo state. When irrigated, yields are similar to those in Florida. Similarly, per hectare yields in humid subtropical regions of China are less than half those of Florida in this instance also due to a lack of irrigation and severe tree losses due to viroids like citrus exocortis viroid (CEV) or in southern China to citrus yellow shoot disease (citrus greening) (see Chapter 6). Climatic factors *per se* are not limiting yields in these areas and there is an enormous potential to increase yields through technological advances, but yields cannot be expected to reach the levels under subtropical conditions where flower bud induction is not limiting.

By comparison, average and maximum yields in semiarid or arid subtropical regions such as California, Spain, Italy, South Africa, Israel and Australia are often lower than those in Florida or irrigated regions of Brazil. In most cases, yields are less because of smaller tree size due to less heat unit accumulation and in some instances due to more intense physiological drop in arid regions. Water quality (especially salinity) and quantity may also limit yields in areas such as Australia. This is not to say, however, that yields are always lower in semiarid or arid subtropical regions than in humid sub-tropical ones. Yields of 80 tonnes ha^{-1} or more for sweet oranges are not uncommon in many of these regions where water is not limiting. Therefore, blanket generalizations concerning yields in various regions are not possible.

Yield potential in tropical regions is also influenced by poorly drained or nutrient-deficient soils in some cases and severe disease and pest pressures in most regions (see Chapter 6). Yields in lowland tropical regions in particular are reduced significantly by the latter problems. Furthermore, trees tend to be excessively vegetative under the temperature and moisture conditions of many low tropical regions. The excessive vigour is not conducive to flower bud induction thus decreasing yields. For example, in tropical areas such as southeastern Mexico, sweet orange yields average only 15–20 tonnes ha^{-1} due primarily to a lack of sufficient irrigation, fertilization and pest control. In contrast, mid-elevation tropical regions like those in Central and South America are potentially as productive as humid subtropical lowland regions if factors like soil moisture are not limiting.

The yield potential of an orchard refers to the maximum amount of fruit mass that is produced per unit of land, but does not consider juice quantity and quality or percentage marketable fruit – a factor of considerable importance to fresh fruit producing countries. Percentage marketable fruit is generally greater in semiarid or arid, subtropical climates such as Spain, California, Italy, etc., than in humid subtropical or tropical locations due to less intense disease and pest pressures, fewer fruit blemishes, and development of more intense and desirable peel colour (see Chapter 7).

ENVIRONMENTAL FACTORS AFFECTING FRUIT GROWTH, DEVELOPMENT AND QUALITY

Climate has a significant effect on fruit growth and quality as clearly demonstrated by Reuther and Rios-Castano (1969) when they compared various fruit quality factors in tropical regions of Colombia with arid and coastal subtropical regions of California. Regions in Colombia were further subdivided into lowland, mid- and highland areas and the California regions into arid and coastal regions differing in average and extreme temperatures, rainfall and humidity.

Fruit Growth

Fruit growth of most citrus cultivars follows a sigmoid pattern which may be subdivided into four phases (Bain, 1958). Phase I is the cell division phase in which nearly all the cells of the mature fruit will be produced. It is this initial cell number that will ultimately determine final fruit size. The duration of this phase ranges from about 1 to 1½ months following bloom depending on climatic conditions and cultivar. During phase II cells differentiate into the various tissue types such as juice sacs, albedo, flavedo, etc. Phase III, the cell enlargement phase, produces a rapid increase in fruit size and percentage total soluble solids (TSS). During this time cells may increase in volume by 1000 times. Phase III duration varies with cultivar from 2 to 3 months for lemons and limes to more than 6 months for sweet oranges and grapefruit. Within a cultivar the duration of phase III may also vary from 3 to 4 months in lowland tropics (Cartagena, Colombia) to 10 months under cool, coastal subtropical conditions (Santa Paula, California).

Peel colour begins to change from green to yellow or orange (except for oranges in lowland tropical regions) towards the end of phase IV, the maturation phase, which is typified by a levelling-off of growth and a slight, gradual increase in TSS along with a rapid decrease in total acidity (TA). The maturation phase may continue for 9–10 months for 'Valencia' oranges under some Mediterranean subtropical conditions, but is one to two months shorter in humid subtropical and considerably shorter yet under lowland tropical conditions. Time from bloom until attainment of an acceptable TSS:TA ratio ranges from 6–7 months in the low tropics to 14–16 months in Mediterranean-type climates for 'Valencia'.

Fruit growth rate within each climatic region is primarily a function of temperature during each developmental stage and soil moisture, particularly during phases III and IV (Fig. 3.7). The highest mean temperatures provide the fastest fruit growth rates (Cartagena, Colombia, lowland tropical) and lowest mean temperatures the slowest (Santa Paula, California, semiarid subtropical, coastal) which is consistent with the heat unit concept discussed

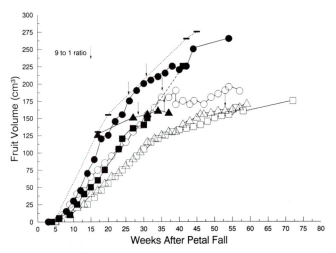

Fig. 3.7. Comparison of growth rates of 'Valencia' oranges in tropical and subtropical locations. The mid-bloom or 'zero' petal fall dates were estimated as follows: △, Riverside, 4 May 1961; ○, Indio, 6 April 1962; □, Santa Paula, 11 May 1964; ▲, Cartagena, about 25 June 1966; ■, Medellín, 20 January 1967. The Palmira data (●) are for the local cultivar 'Lerma' which appears to be a mid-season type. Source: Reuther and Rios-Castano (1969).

previously (Table 3.1). However, adequate soil moisture via rainfall or irrigation significantly improves fruit size during phase III but a corresponding dilution of soluble solids results (see the projected fruit growth rate [dotted line] for Cartagena). Fruit growth rates are intermediate and similar for high tropics (Medellín, Colombia) and arid subtropical (Indio, California) regions. Fruit growth rates for humid subtropical regions like São Paulo, Brazil, or Orlando, Florida, are similar to those of Palmira, Colombia.

External Quality

External quality factors including peel colour, incidence of blemishes and fruit shape are significantly affected by climate. Peel colour of citrus fruits results from a combination of pigments including chlorophyll, carotenoids, antho-cyanins and lycopene among others. Initially the peel cells contain high levels of chlorophyll allowing the fruit to produce some photosynthetic metabolites. Typically the peel provides no more than 10% of the total carbon require-ments of the young fruit and much less for the mature fruit, with the remainder being imported from the leaves. In growing regions where the average temperatures remain high all year (e.g. lowland tropical regions), chlorophyll levels remain high for oranges and mandarins and the fruit peel stays green. However, as air and soil temperatures fall below 15°C, chlorophyll is degraded and chloroplasts are converted to chromoplasts containing yellow, orange or red pigments (carotenoids, lycopenes, etc.). Carotenoid synthesis is reduced above 35°C or below 15°C but still occurs at temperatures conducive to chlorophyll degradation. Peel colour in grapefruit results from yellow carotenoids and red lycopene synthesis in pink and red cultivars (Erickson, 1968). Lycopene synthesis proceeds even at moderate average temperatures, but synthesis is retarded at high temperatures. Grapefruit attain a yellowish to red–orange peel colour, depending on cultivar, even in low tropical regions after slow chlorophyll breakdown. Of interest, the peel of certain cultivars like Valencia orange can undergo reversion from orange to green, a process termed 'regreening'. In this instance, chromoplasts revert to chloroplasts. Regreening generally occurs in late-harvested cultivars after fruits have remained on the tree late into their normal harvest period and spring high temperatures and adequate soil moisture result in stimulation of chlorophyll synthesis. The blood oranges, popular in Italy (Chapter 1), contain anthocyanin pigment and require prolonged cool temperatures for good colour development.

Besides climate (temperature), tree vigour also has a pronounced effect on fruit colour. Generally, vigorously growing trees produce more poorly coloured fruit than slower-growing trees. As a result, any factor that enhances vigour delays the development of peel colour. Fruit from young vigorous trees or from those on vigorous lemon-type rootstocks are generally more poorly coloured than those from slower growing trees within a particular growing region. Further, trees receiving excessive nitrogen tend to have poor peel colour. As light is necessary for carotenoid and anthocyanin synthesis, shaded fruit will be more poorly coloured than exposed fruit. Orchards that have trees grown into hedgerows produce few and poorly coloured fruit toward the interior of the tree.

Blemishes are major factors that prevent citrus fruits from being marketed fresh. Blemishes are produced due to abiotic and biotic factors (see Chapter 7). Incidence of blemishes caused by biotic factors is usually much greater in humid subtropical and tropical regions than in semiarid or arid subtropical regions due to increased pest and disease pressures. Insect, mite and fungus diseases may be extremely intense and difficult to control in humid lowland areas of the tropics (see Chapter 6).

Fruit shape is adversely affected in grapefruit by high temperatures during the cell division stage. Excessive cell division in the albedo near the stem end produces an elongated, 'sheepnosed' fruit.

Internal Quality

Water comprises a major portion of the fruit mass (85–90% by weight), with carbohydrates contributing to 75–80% of the TSS (see Chapter 7). As a

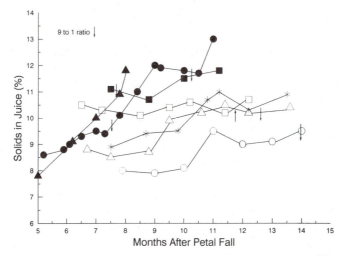

Fig. 3.8. Comparison of trends in percent total soluble solids in the juice of 'Valencia' oranges grown in Colombia and California. Symbols: ▲, Cartagena; ●, Palmira; ■, Medellín; *, Indio; □, Lindsay; △, Riverside; ○, Santa Paula. Source: Reuther and Rios-Castano (1969).

consequence, regulation of carbohydrate loading into the citrus fruit has a great impact on internal fruit quality. Fruit growth is a function of tree water status and carbohydrate partitioning as well as temperature. Fruit shrink and swell diurnally as the water relations of the tree change. Ultimate fruit size is increased by irrigation and/or rainfall. The fruit in addition serve as a storage organ for water. Leaves on a detached fruitless twig wilt within hours, while those on a detached fruited twig remain turgid for many hours. A majority of the water translocated to the leaves is stored in the fruit peel. Changes in fruit volume have been used to schedule irrigation in some areas such as Australia, California and Arizona. This practice, while quite reliable, is not widely used due to the amount of labour involved (Hilgeman, 1977).

Total soluble solids accumulate most rapidly in the fruit under lowland tropical conditions and most slowly under cool coastal conditions (Fig. 3.8). Maximum levels of TSS are usually attained in the midtropics (Palmira, Colombia) and in humid subtropical regions with warm winters such as São Paulo, Brazil, or Florida. Levels of TSS are intermediate in semiarid or arid subtropical and highland tropical areas such as Riverside, California, or Medellín, Colombia.

Of more importance to attainment of commercial edibility of citrus fruit than TSS, is the rate of decline in TA. Under hot lowland tropical conditions such as those in Cartagena, Colombia, TA decreases rapidly from 2.0% (for sweet oranges) to below 0.5% (Fig. 3.9). Orange juice from fruit in these areas is often insipid because of the lack of acidity. However, grapefruit attain very high quality due to reduced TA in hot, tropical regions. A similar but more

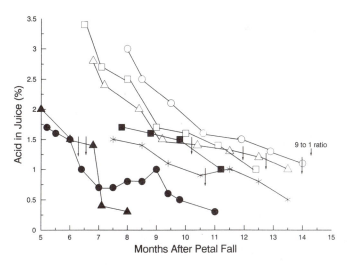

Fig. 3.9. Comparison of the trends in total acid (calculated as anhydrous citric) concentration in the juice of 'Valencia' oranges in relation to advancing maturity. Symbols: ▲, Cartagena; ●, Palmira; ■, Medellín; *, Indio; □, Lindsay; △, Riverside; ○, Santa Paula. Source: Reuther and Rios-Castano (1969).

protracted decrease in TA occurs in mid-elevation tropical and humid subtropical areas. Juice may also become insipid in these regions if fruit are held on the tree for extended periods. Total acid levels are generally greatest in semiarid or arid subtropical and coastal regions and decline more slowly than in other regions. This decrease in TA is primarily a function of temperature (heat unit accumulation) and the rapid respiration of organic acids at these temperatures, although rate of decline is also increased and minimum levels decreased by excessive rainfall or irrigation, or by selection of certain rootstocks (Chapter 4).

The ratio of TSS:TA is a primary determinant of fruit edibility and is linked to maturity standards in many citrus-growing regions (see Chapter 7). For example, attainment of a minimum TSS:TA ratio of 9:1 for 'Valencia' sweet oranges occur in 6–7 months in lowland or midtropical areas, may not occur for 8–12 months in upland tropical and subtropical areas, and is slowest in coastal subtropical areas, taking as long as 14–16 months (Fig. 3.10). These time frames, of course, change with cultivar, rootstock and value for minimum ratio. However, the general trends are related to climate irrespective of cultivar.

It is quite obvious that climate has a significant effect on nearly all aspects of citrus growth and development. Failure to assess accurately the impact of climate on economic profitability of citrus is a major reason for crop losses (in

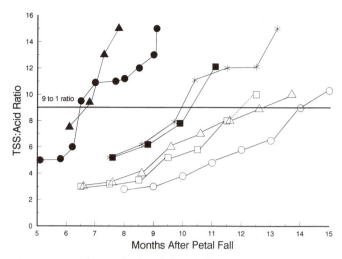

Fig. 3.10. Comparison of the trends of ratio of total soluble solids and total acid concentrations in the juice of 'Valencia' oranges in relation to advancing maturity in tropical and subtropical climates: Symbols: ▲, Cartagena; ●, Palmira; ■, Medellín; ✳, Indio; □, Lindsay; △, Riverside; ○, Santa Paula; -------, extrapolation. Source: Reuther and Rios-Castano (1969).

the case of severe freeze-damage) or at least reduced income because of low yield and quality potential. It is important to note that climatic factors are both limiting and coactive. For example, in most citrus regions light intensity is not a limiting factor for tree growth but low light may reduce yields especially in the interior of the canopy. Low temperature and water are certainly major limiting factors but temperature and water stress may also act together to reduce or enhance growth or productivity. Therefore, a successful citrus grower must first choose a location that ultimately optimizes climatic factors that are coactive and then reduce or limit the risk of limiting factors. Therefore, choice of a site with optimum temperature conditions for a particular cultivar is of primary concern. Factors such as irrigation and drainage can then be regulated to achieve optimum returns.

4

ROOTSTOCKS

For hundreds of years citrus trees were grown as seedlings. Seedling trees, however, usually have a protracted juvenile period. Juvenility is undesirable in most citruses because trees are nonproductive and excessively vigorous, resulting in upright growth with thorny branches. A protracted juvenility period is particularly undesirable in the 1990s where the emphasis has been on producing a crop and returning an investment as soon as possible after planting. Most seedling trees are also susceptible to several soil-related problems, in particular *Phytophthora parasitica* and burrowing and citrus nematodes, and may not be true-to-type for some cultivars with low rates of nucellar embryony (see Chapter 5).

Consequently, most citrus orchards worldwide consist of two-part (usually budded) trees that combine favourable attributes of the scion and rootstock. Rootstock selection is a major consideration in every citrus-growing operation. It is fundamental to the success of the orchard because the rootstock chosen will become the root system of the budded tree. Besides supporting the tree, the root system is responsible for absorption of water and nutrients, providing storage of carbohydrates produced in the leaves and synthesis of certain growth regulators, adapting the scion to particular soil conditions, and potentially providing tolerance to some diseases. More than 20 horticultural characteristics are influenced by the rootstock including tree vigour and size, depth of rooting, freeze tolerance, adaptation to certain soil conditions, such as high salinity or pH, or excess water, resistance or tolerance to nematodes and diseases like phytophthora foot rot and citrus blight, and fruit yield, size, texture, internal quality and maturity date (Castle, 1987). Some problems like sensitivity to excess copper or damage due to root weevils or armillaria root rot cannot currently be remedied by choice of rootstock.

There is no perfect rootstock, even for a particular situation. Choice of rootstock should be based on the most important limiting factor(s) to production in a particular region, local climate and soil conditions, cultivar and intended use (fresh or processed) of the crop. For example, sour orange

Table 4.1. Characteristics of important rootstocks of citrus.

Characteristics	Rough lemon	'Milam' lemon	'Volkamer' lemon	Citrus macrophylla	'Rangpur' lime	'Palestine' sweet lime	Sour orange
Salinity tolerance	I	~	~	G	G	P	I
High calcium tolerance	H	~	H	H	H	I	H
Tree vigour	H	H	H	H	H	H	I
Fruit size	LG	LG	LG	LG	LG	LG	I
Brix	L	L	L	L	L	L	H
Yield/tree	H	H	H	H	H	H	I
Citrus nematode tolerance	S	S	S	S	S	S	S
Burrowing nematode tolerance	S	R	S	S	S	S	S
Xyloporosis tolerance	T	T	T	S	S	S	T
Exocortis tolerance	T	T	T	T	S	S	T
Tristeza tolerance	T	T	T	S	T	T	S
Blight tolerance	P	P	P	P	P	P	G
Freeze tolerance	P	P	P	P	P	P	G
Drought tolerance	G	G	G	G	G	G	I
Flood tolerance	G	G	G	G	~	G	I
Phytophthora tolerance	S	S	S	R	S	S	T

'Smooth Flat Seville'	T	I	I	G	G	S	T	T	S	S	I	H	I	I	H	—
'Cleopatra' mandarin	T	P	I	G	G	T	T	T	S	S	LI	H	SM	H	I	G
Sweet orange	S	P	P	I	G	T	T	T	S	S	I	I	I	L	L	—
Ridge pineapple	S	P	P	I	G	T	T	T	R	S	I	I	I	L	L	—
Trifoliate orange	R	I	P	G	P	S	S	T	S	R	LI	H	SM	L	L	P
'Carrizo' citrange	T	P	G	I	P	S	S	T	T	T	H	I	I	H	L	P
'Troyer' citrange	R	G	I	G	I?	T	T	T	S	R	I	H	I	I	I?	—
'Swingle' citrumelo	T	?	G	I	?	T	S	T	S	T	H	I	I	H	L	P
Grapefruit	S	?	P	I	?	T	T	T	S	S	H	L	SM	H	L	—

Adapted from: Castle *et al.* (1987).

Key to symbols: G = good; H = high; I = intermediate; L = low; LG = large; P = poor; R = resistant; S = susceptible; SM = small; T = tolerant; ? = inadequate information or rating unknown.

should not be used as a rootstock in areas such as Spain, Brazil, California, South Africa or the south Caribbean where citrus tristeza virus (CTV) is prevalent and will kill or seriously debilitate trees. Rootstocks which import excessive vigour to the scion, like rough lemon or 'Palestine' sweet lime, should be avoided in areas like northern Florida and Texas (USA) that are susceptible to regular freeze-damage and rough lemon rootstock should be avoided in areas known to have blight. Scions budded on *Poncirus trifoliata* and most citranges and citrumelos perform poorly in high pH soils. There are many other examples which will be discussed in the following sections further demonstrating that identification of the most limiting factors in a given area related to rootstock is the first and probably most important step in choosing a rootstock.

Local climatic and soil conditions are a first consideration in rootstock selection. Sour orange is used almost exclusively in Texas due to its adaptability to alkaline and saline soils. In contrast rough lemon, with its excellent drought tolerrance, is widely planted in the sandy soils of South Africa and Australia. *Poncirus trifoliata* is well-adapted to the cool growing conditions and acid soils found in Japan and central China and 'Carrizo' citrange is an important rootstock where the burrowing nematodes are a problem. Consequently, more citrus regions have limited their rootstock selections based on local conditions and traditions, a factor that is potentially very dangerous.

Cultivar and intended use (fresh or processing) are also important for rootstock selection. 'Cleopatra' mandarin ('Cleo') is well-suited for use with mandarins and mandarin-hybrids ('Temple', 'Robinson', 'Nova', 'Sunburst'), but generally sweet orange and grapefruit cultivars on 'Cleo' produce small fruit, and are not precocious. 'Cleo', however, provides CTV tolerance and moderate blight tolerance. Rootstocks that impart high vigour to the scion (e.g. rough lemon, *Citrus volkameriana*, 'Palestine' sweet lime, and 'Rangpur' lime) generally induce high yields, but produce relatively poor quality fruit, low in total soluble solids and acids (TSS), with coarse peels, more acceptable for processing than for the fresh market (Castle, 1987). Conversely, grapefruit and sweet orange trees on sour orange, 'Troyer' and 'Carrizo' citranges and 'Swingle' citrumelo rootstocks typically produce high quality fruit suitable for the fresh and processing markets.

Important horticultural characteristics of the major rootstocks worldwide are discussed in the following section in more detail and summarized in Table 4.1 (Castle, 1987). Ratings given in the table are based on years of observations and controlled comparisons of rootstock characteristics; however, these ratings are not representative of every growing condition worldwide. Climatic conditions, cultivar and soil characteristics in particular may alter some of the horticultural characteristics of each rootstock. For additional information on citrus rootstocks refer to reviews by Wutscher (1979) and Castle (1987).

DESCRIPTION OF MAJOR CITRUS ROOTSTOCKS

Rough Lemon

Rough lemon (*Citrus jambhiri* Lush.), which is native to northeastern India, is probably a natural hybrid because of its high degree of polyembryony compared with other lemon species (Barrett and Rhodes, 1976). The fruit has, as the name implies, a very coarse exterior and rough lemon is unsuitable as a scion cultivar; however, it has been widely used in many countries as a rootstock. The adaptability of the species as a rootstock for deep infertile sands has prompted its use in the ridge (central) area of Florida, Australia and South Africa. However, the high susceptibility of rough lemon to blight and its lack of freeze-tolerance (Florida) has virtually eliminated its use in Florida and Brazil.

Sweet orange, grapefruit, mandarin and lemon trees on rough lemon rootstock are large, extremely vigorous and very productive in most rootstock trials worldwide, particularly in deep, sandy soils (see Castle, 1987 for references). Yields of mature grapefruit trees in areas like Florida have attained 80–100 tonnes ha^{-1} in some instances. The root system is extensive, sometimes reaching a depth of 4.6 m in deep sands (Castle and Krezdorn, 1973). Consequently, mature trees on this rootstock are also very drought tolerant. Rough lemon rootstock is moderately tolerant of high salinity. In addition, it is well-adapted to a wide range of soil pHs.

Scions on rough lemon are very susceptible to freeze-damage due to their vigorous growth over a wide temperature range (Yelenosky and Young, 1977). Thus, the scion often does not become fully freeze-acclimated, possibly because of the relatively higher root conductivity of rough lemon at lower root temperatures than found in 'Carrizo' citrange for example (Wilcox and Davies, 1981). Scions on rough lemon-type rootstocks have been consistently more damaged by freezes than those on other rootstocks if soil temperatures remain high, >15°C, and trees are poorly freeze-acclimated. In Mediterranean-type climates where winter soil temperatures are low, citrus trees on rough lemon became moderately freeze-hardy. Regrowth of scions on rough lemon following freeze-damage is usually very vigorous and thus the tree recovers more rapidly than on other rootstocks.

Rough lemon is tolerant to CTV, citrus exocortis viroid (CEV) and citrus xyloporosis viroid, although some dwarfing may occur in scions infected with the last two diseases (Wutscher, 1979). There is also some evidence from South Africa, however, that severe strains of CTV may cause stem pitting even on rough lemon. It is highly susceptible to foot rot (*Phytophthora*), a factor which limits its use in many regions, and it is susceptible to damage by both the citrus and burrowing nematodes (O'Bannon and Ford, 1977). 'Milam', a probable rough lemon hybrid, produces similar characteristics in the scion as rough lemon but is resistant to the burrowing nematode. Most importantly, rough lemon is highly susceptible to blight, a disease of unknown cause responsible for extensive tree losses, particularly in Florida (Young *et al.*, 1982).

The excessive vigour imparted to the scion by rough lemon generally produces poor quality fruit. Both TSS and titratable acidity (TA) tend to be low in fruit of trees on rough lemon (Castle, 1987). For example, TSS for 'Valencia' on sour orange averaged 13.2% while on rough lemon TSS averaged 11.4% (Gardner and Horanic, 1961). Moreover, the peel tends to be thick and puffy, especially for fruit on vigorously growing young trees. Mandarin fruit in particular for trees grown on rough lemon rootstock tends to be puffy and hold poorly on the tree. Regreening of 'Valencia' oranges is more severe on rough lemon than on other rootstocks. Fruit size is generally quite large for all cultivars grown on rough lemon and use of rough lemon for inherently small-fruited cultivars like Dancy mandarin may be advantageous in some situations, but granulation tends to be more severe on rough lemon. In many citrus regions fruit produced on rough lemon does not attain sufficient quality for the export fresh fruit market but total kg solids production per tree and hectare are higher for scions budded on rough lemon than for most other rootstocks (Castle, 1987). The high production of kg-solids (termed pounds-solids in Florida) per hectare also produces high returns for the processing market. Total kg-solids per hectare is a function of yield, juice content and TSS.

Citrus volkameriana

Citrus volkameriana Ten. and Pasq. ('Volkamer' lemon) is also a lemon hybrid which as a rootstock produces large, vigorous trees yielding large quantities of moderate to poor quality fruit like rough lemon (Castle, 1987). Scions on 'Volkamer' lemon are slightly more freeze-hardy than those on other lemon-types. They are not susceptible to CTV, xyloporosis or CEV, but are susceptible to blight and the citrus and burrowing nematodes (W.S. Castle, unpublished). However, they are tolerant to malsecco and phytophthora under most circumstances (Carpenter *et al.*, 1981).

Volkameriana is not widely used as a rootstock and probably will not come into widespread use in the near future, although it has been planted extensively in recent years in Venezuela and other Caribbean basin countries which have lost sour orange rooted trees to CTV. Unfortunately, 6-year-old trees are now being lost to citrus blight. Nevertheless, some studies from Florida suggest that yields and net profits over the long-term are higher for trees on 'Volkamer' than for other less vigorous rootstocks such as 'Swingle' citrumelo even when tree losses due to blight or other factors are moderate (W.S. Castle, unpublished).

Citrus macrophylla (Alemow)

Citrus macrophylla Wester is a hybrid species, possibly of *Citrus celebica* and *Citrus grandis*, native to the Philippines (Barrett and Rhodes, 1976). It is very similar morphologically and genetically to lemons and limes. Cultivars budded on *C. macrophylla* produce large, vigorous and high-yielding trees with growth characteristics similar to those on other lemon-type rootstocks under most growing conditions (Castle, 1987). Carpenter *et al.* (1981), in contrast, found 'Eureka' lemon trees on *C. macrophylla* were smaller than those on 'Swingle' or *C. volkameriana*. Fallahi and Rodney (1992) observed that 'Fairchild' mandarin trees were more precocious on *C. macrophylla* than on *C. volkameriana* or 'Carrizo' citrange. Moreover, yields of 'Valencia' orange on *C. macrophylla* were similar to, or better than, those on other lemon-types but superior to those produced on trifoliate and trifoliate hybrids (Fallahi and Rodney, 1992; Castle, 1987). Lemons and limes also yield very well on this rootstock under most situations (Castle, 1987). Scions budded on *C. macrophylla* grow well on both sandy and high pH, calcareous soils. Trees on *C. macrophylla* have a deep, dense root system that imparts drought tolerance to the scion.

Citrus macrophylla itself and scions budded to it are freeze-sensitive. Scions budded on *C. macrophylla* are more freeze-tender than those on rough lemon and are far less freeze-hardy than scions on sour orange, 'Cleopatra' mandarin, trifoliate orange or 'Swingle' citrumelo (Castle, 1987). As with rough lemon, regrowth is very rapid following moderate freeze-damage.

Scions on *C. macrophylla* (except lemons) are susceptible to CTV and xyloporosis, although to a lesser extent than sour orange or 'Palestine' sweet lime, respectively (Castle, 1987). *C. macrophylla* is more tolerant to foot rot than true lemons (Carpenter *et al.*, 1981), but is not tolerant of citrus or burrowing nematodes, and is moderately susceptible to blight (Castle, 1987).

Fruits of sweet oranges and grapefruit budded on *C. macrophylla* are generally large. They may become over large and puffy especially on young trees. Lemon and lime fruits grown on this rootstock are also large, which is a favourable characteristic. Fruit quality is moderate to poor, similar or slightly better than that on other lemon-type rootstocks. Generally, TSS, TA and TSS:TA ratio are lower on this rootstock than on sour orange, 'Carrizo' citrange or 'Swingle' citrumelo (W.S. Castle, unpublished).

C. macrophylla is an excellent rootstock for lemons and limes but is not widely used for sweet oranges, grapefruit and mandarins. However, 'Eureka' lemon trees budded on *C. macrophylla* may develop a disorder called *C. macrophylla* rootstock necrosis. Blockage of the sieve elements leads to necrosis and eventually tree decline and death (Schneider *et al.*, 1978). There is evidence that *C. macrophylla* is better adapted to cool, dry climates. When used as a rootstock in such climates, e.g. Spain, many cultivars on *C. macrophylla* out-yield those on citranges.

'Rangpur'

'Rangpur' (*Citrus reticulata* hybrid) is a mandarin-type hybrid (Wutscher, 1979). 'Rangpur' is not widely planted in most citrus areas with the exception of Brazil where it is the most important rootstock primarily because of its tolerance to CTV and drought. Cultivars budded on 'Rangpur' are moderately vigorous and yields are similar or slightly lower than those produced on rough lemon for lemons (Castle, 1987) and greater than those found on most citranges, trifoliata and 'Cleopatra' mandarin rootstocks. 'Rangpur' is tolerant of high saline and calcareous soil conditions.

Fruit quality and freeze-tolerance are intermediate to those of lemon-types and sour orange. 'Rangpur' is resistant to CTV, but susceptible to CEV, the citrus and burrowing nematodes and moderately susceptible to foot rot. Observations in Brazil show that 'Rangpur' is very susceptible to blight (Lima, 1982). Although sweet orange trees on 'Rangpur' have a number of favourable characteristics it is unlikely that it will become an important rootstock outside of Brazil.

'Palestine' Sweet Lime

'Palestine' sweet lime (*Citrus limettoides* Tan.), which is probably a hybrid rather than a true lime, is of minor importance worldwide as a rootstock (Castle, 1987). Cultivars budded on sweet lime are quite similar in vigour, yields and fruit quality characteristics to rough lemon and marginally superior to lemon-types in freeze-tolerance. Trees on 'Palestine' are moderately susceptible to CTV and foot rot, and may be stunted by CEV and particularly xyloporosis (Wutscher, 1979). In fact, 'Palestine' sweet lime, formerly a major rootstock in Israel, has been largely replaced due to its susceptibility to xyloporosis. Trees on 'Palestine' sweet lime are also susceptible to blight.

Sour Orange

Sour orange (*Citrus aurantium* L.) has been and probably continues to be the most widely planted rootstock in the world as of 1993. However, susceptibility of sweet orange on sour orange to CTV has greatly decreased its use for new plantings in Australia, Argentina, Brazil, California, Spain, South Africa, most of Florida and southern Caribbean countries. Sour orange is an excellent rootstock for areas free of CTV and particularly for fresh fruit production (Castle, 1987).

Cultivars grown on sour orange produce trees of moderate vigour and moderate to large size (Hutchison, 1977). Trees on sour orange grow somewhat more slowly than those on rough lemon-types but are certainly not dwarfed. Yields on a mass or kg-solids basis are also adequate, but less than those on rough lemon (Hutchison, 1977), *C. macrophylla* or *C. volkameriana* on sandy soils (Castle, 1987).

Sour orange produces a deep and moderately branched root system and scions budded on this rootstock are moderately drought-tolerant. Scion vigour on deep sandy soils is certainly not as good as on more fertile soils where water and mineral elements are not limiting. Sour orange rootstock is commonly used in heavy or poorly drained soils due to its moderate phytophthora tolerance, although it is not as physiologically tolerant of flooding as rough lemon rootstock (Syvertsen *et al.*, 1983). Trees on sour orange rootstock are particularly well-adapted to high pH and high salinity soils (Wutscher, 1979).

Cultivars budded on sour orange are about as freeze-hardy as those growing on 'Cleopatra' mandarin, *P. trifoliata*, or 'Swingle' citrumelo when fully acclimated, but are appreciably superior to lemon-types and citranges (Yelenosky and Young, 1977). Regrowth of freeze-damaged scions, however, is slower for trees on sour orange compared with those on rough lemon or other vigorous rootstocks (A.H. Krezdorn, unpublished).

Tree stunting, bark sloughing or stem pitting commonly associated with CEV and xyloporosis do not occur when sour orange is used as a rootstock. Phytophthora foot rot is usually a minor problem for trees on sour orange rootstock. Nevertheless, in some countries such as Mexico and Spain trees are budded high to avoid foot rot problems, and sour orange trees are moderately susceptible to root rot. Sour orange is susceptible to damage by the burrowing and citrus nematodes (O'Bannon and Ford, 1977) but is one of the rootstock's least susceptible to blight (Young *et al.*, 1982).

Fruit size of cultivars on sour orange is somewhat smaller than on rough lemon, but larger than on 'Cleopatra' mandarin. Both the TSS and TA of the juice are high, thus sour orange has been the preferred rootstock in fresh fruit-producing regions such as Spain, the Indian River area of Florida, and Texas (Castle, 1987). High TA frequently results in fruit meeting maturity standards later than fruit from trees on more vigorous rootstocks. In contrast, in certain cultivars like Hamlin orange for which the limiting maturity factor is low TSS, fruit will attain earlier maturity on sour orange than on a rootstock like rough lemon. Total soluble solids for fruit from scions on sour orange often average 0.5 to 1.5% higher than for those on rough lemon. This characteristic along with moderate phytophthora tolerance has made sour orange the preferred rootstock in tropical countries where CTV is not limiting.

The peel of fruit from cultivars grown on sour orange generally is smooth and thin. For this reason, excessive splitting can occur, although the problem is less severe than for the same cultivars on 'Cleopatra' rootstock (A.H. Krezdorn, unpublished). While splitting can be a serious problem with sweet oranges and some mandarins or mandarin-hybrids, it is not important in either grapefruit or 'Temple' oranges and generally is not severe enough to limit its use as a rootstock.

Recently there has been interest in using selections of sour orange such as 'Bittersweet' and *Citrus taiwanica*. 'Bittersweet' has been available as a rootstock for many years. Studies from the 1960s in Florida suggest that most sour orange selections have similar effects on scion fruit quality and performance and therefore there appears to be no advantage to using one type of sour orange over another. 'Bittersweet' is not tolerant of CTV but may be more tolerant of phytophthora than sour orange and *C. taiwanica* was thought to be CTV-tolerant, although this has not been substantiated (Castle, 1987). Wutscher (1977), in summarizing 30 years of rootstock studies with red grapefruit in Texas, observed slightly higher yields for 'Bittersweet' over *C. taiwanica* and sour orange.

'Smooth Flat Seville' (Australian Sour)

'Smooth Flat Seville' (pronounced se'vil) which originated in Australia, is probably a hybrid of pummelo, sweet orange and sour orange (Barrett and Rhodes, 1976). Its characteristics have not been widely compared to those of other rootstocks and 'Seville' has been tested as a rootstock in only a few experimental plantings. Yields and vigour of trees budded on this rootstock are moderate for sweet oranges and moderate to good for grapefruit. Fruit quality is slightly poorer than on sour or trifoliate orange but superior to that on lemon types. Trees on 'Seville' are not susceptible to xyloporosis and CEV and are moderately susceptible to CTV and phytophthora, although there is some question about its tolerance to these diseases. 'Smooth Flat Seville' is also susceptible to citrus and burrowing nematode damage. Scions budded on 'Seville' appear to have good blight tolerance.

'Cleopatra' Mandarin

'Cleopatra' mandarin (*C. reticulata* Blanco) is of minor importance as a rootstock on a worldwide basis; however, it has several favourable attributes which have increased its use in recent years. Scions budded on 'Cleopatra' mandarin are large and moderately vigorous (Hearn and Hutchison, 1977), producing a deep, densely branched root system that imparts moderate drought tolerance. 'Valencia' and 'Parson Brown' sweet oranges on 'Cleopatra' rootstock had moderate yields during a 17-year study in Florida (Gardner and Horanic, 1961), being less than those for trees on rough lemon but greater than those on sour orange rootstock. Similarly, yields of 'Ruby Red' grapefruit were comparable with those on sour orange but considerably lower than those on rough lemon. Moderate yields resulted from poor fruit set and size and splitting of mature fruit (A.H. Krezdorn, unpublished). Scions on 'Cleo' are not precocious, which is a major factor limiting the selection of 'Cleo' as a rootstock, but attain moderately large size and yields 10–15 years after planting (Gardner and Horanic, 1961). In south Florida, recent improvements in irrigation and fertilization practices have resulted in faster growth and attainment of moderate yields at an earlier age for trees on 'Cleopatra' mandarin rootstock (see Chapter 5).

Trees on 'Cleopatra' mandarin are as freeze-hardy as those on sour orange, *P. trifoliata*, or 'Swingle' citrumelo when fully acclimated, but are appreciably superior to trees on rough lemon or 'Carrizo' citrange (Yelenosky and Young, 1977). Regrowth following freeze-damage is less than that of rough lemon but comparable to that observed for sour orange.

A major advantage of 'Cleo' over many other rootstocks is its tolerance of the major citrus virus (viroid) diseases (Wutscher, 1979). 'Cleopatra' mandarin is tolerant of CTV, CEV and xyloporosis, displaying none of the typical symptoms associated with these problems. There are observations that 'Cleopatra' is not completely tolerant of xyloporosis, but such cases seem to be exceptions. 'Cleopatra' is susceptible to both burrowing and citrus nematode damage and has moderate phytophthora foot rot and poor root rot tolerance because damaged roots regrow very slowly. 'Cleopatra' usually reaches an age of 12–15 years before losses to blight occur, but after this age blight losses can become fairly high (Young *et al.*, 1982).

'Cleopatra' mandarin rootstock is also adapted to a wide variety of soils ranging from light sands to heavy clays, although scions budded on it are most productive on heavier soils. It is resistant to high salinity and is tolerant of high pH, calcareous soils.

Fruit size of cultivars on this rootstock is consistently smaller than that of trees on other commercially important rootstocks (A.H. Krezdorn, unpublished). Juice of fruit produced on 'Cleopatra' is of moderately high quality. Total soluble solids are usually intermediate between those of sour orange and rough lemon (Castle, 1987), although Economides (1976) found that TSS was higher for 'Marsh' grapefruit on 'Cleo' than on sour orange and rough lemon in Cyprus. The peel is smooth and thin, which is apparently related to the excessive splitting commonly found in fruit produced on 'Cleopatra'. Splitting can be a serious problem with sweet oranges in some years, but is unimportant for grapefruit.

'Cleopatra' is not widely used as a rootstock for sweet oranges and grapefruit, or for some small-fruited mandarins like 'Dancy'. In contrast, 'Cleopatra' is an excellent rootstock for 'Temple' and is widely used for mandarin hybrids such as 'Orlando', 'Nova', 'Murcott' ('Honey'), 'Robinson', 'Sunburst' and 'Minneola' (Hearn and Hutchison, 1977). The size and quality of the fruit of these cultivars on 'Cleopatra' are excellent; however, with the exception of the self-fruitful 'Murcott' and 'Temple', the yields are low unless adequate cross-pollination with a compatible cultivar is provided.

Although sweet oranges and grapefruit are not precocious when budded on 'Cleo', tree survival and longevity are usually very good, particularly in areas where blight and CTV are prevalent. Brazilian and Venezuelan growers have been planting 'Cleo' as an alternative to more blight susceptible rootstocks since sour orange cannot be used because of its sensitivity to CTV.

Sweet Orange

Interest in using sweet orange (*Citrus sinensis* [L.] Osb.) as a rootstock is based on favourable performance in California in the past, its moderate to high tolerance to citrus blight and CTV (Young *et al.*, 1982), and the possibility of regrowing trees from their own roots after freezes. Sweet orange-rooted trees are not precocious but eventually become vigorous and moderately productive. Gardner and Horanic (1961) observed that yields of 'Valencia' sweet oranges on sweet orange were similar to those on 'Cleopatra' mandarin, sour orange and grapefruit, but considerably less than those on rough lemon rootstock, and fruit quality is intermediate between lemon-types and sour orange rootstocks. Sweet orange has not been commonly used as a rootstock in most of the world except in Brazil and Australia because of limited drought tolerance and extreme foot rot susceptibility (Castle, 1987). Nevertheless, use of improved microirrigation systems and systemic fungicides may permit expanded use of sweet orange in the future. In limited tests in Florida and Texas own-rooted cuttings of sweet orange have performed fairly well, although long-term data are lacking. Scions budded on sweet orange are not susceptible to CEV or xyloporosis, but most selections are adversely affected by burrowing and citrus nematodes (O'Bannon and Ford, 1977).

Trifoliate Orange

The trifoliate orange (*P. trifoliata* [L.] Raf.) is widely used as a rootstock for satsuma mandarins and sweet oranges in some areas of the world such as Japan, China, Argentina and Australia. Trifoliate orange differs from other rootstocks because many selections are available that have variable effects on scion characteristics. This fact probably accounts for variations in yields and disease responses reported for scions budded on trifoliate orange. For many years trifoliate orange was considered a dwarfing rootstock. Dwarfing of scions on this rootstock, however, is primarily due to infection with CEV (Cohen, 1968), although certainly most scion cultivars budded on trifoliate orange are clearly less vigorous than those on lemon-type rootstocks. Moreover, some strains of CEV do not cause bark scaling but still dwarf the tree. In fact, inoculation with CEV has been used commercially to dwarf trees on trifoliate orange in countries such as Israel and Australia (Bevington and

Bacon, 1977). Yields of most scion cultivars budded on trifoliata are less than those on rough lemon, 'Rangpur' and sour orange in tropical and subtropical areas, primarily due to differences in tree vigour (Castle, 1987). Nevertheless, yields of satsuma mandarin on trifoliata are moderate to good in cooler growing regions such as Japan and central China where yields approach 80 tonnes ha^{-1}.

Trifoliate orange itself is deciduous, becoming very dormant and extremely freeze-hardy. Trifoliate orange trees survive as far north as Long Island, New York (USA) (42°N latitude). This has led to the common misconception that trees budded on trifoliate orange are always much more freeze-hardy than those on other rootstocks. In subtropical regions such as Florida trees become quiescent slightly later when on this rootstock than when budded on 'Cleopatra' mandarin or sour orange (Young, 1977). Thus, early in the winter scions budded on trifoliate orange may be slightly more sensitive to freeze-damage than when on the other two rootstocks (Young, 1977). Moreover, Yelenosky and Hearn (1967) observed that a series of freezes in a given year had a more detrimental effect on scions on trifoliate than on rough lemon-type rootstocks. In contrast, during periods of cool night-time temperatures (<10°C), trees on trifoliate orange cease root growth and became fully quiescent and equal or greater in hardiness to trees on sour orange, 'Cleopatra' mandarin and 'Swingle' citrumelo.

Trees budded on *P. trifoliata* are not affected by CTV or xyloporosis, but are susceptible to blight. Young *et al.* (1982) found less spread of blight for citrus trees on trifoliata than those on rough lemon, but orchard-to-orchard variability was very high in the limited numbers of orchards sampled. However, plantings of trifoliate in Florida and Brazil are not widespread enough to make conclusive statements about its blight susceptibility. Trifoliate orange in general is resistant to the citrus nematode but not to the burrowing nematode (O'Bannon and Ford, 1977). Not all selections of trifoliate orange, however, are resistant to the citrus nematode. Trifoliate orange is highly resistant to foot rot and other problems associated with poorly drained soils.

Scion cultivars budded on trifoliate orange grow poorly on infertile, sandy soils and are not drought-tolerant but grow quite well on moderately fertile sands; trifoliate orange is better adapted than most rootstocks to heavy, poorly drained soils. Trifoliate orange is not well adapted to high salinity and high pH, calcareous soils, however (Von Staden and Oberholzer, 1977). Leaves of scions budded on to trifoliate become quite chlorotic in high pH soils and growth and yields are decreased.

Fruit size of cultivars budded on trifoliate orange varies with soil type. Some studies indicate that fruit of trees on trifoliate orange is exceptionally large, but in general trees on this rootstock bear relatively small fruit (A. H. Krezdorn, unpublished). This may result from the fact that trifoliate sets large crops of fruit even under drought conditions with a concomitant reduction in fruit size. The juice of fruit produced on trifoliate orange is of excellent quality, rating as good or better than that on any other rootstock (Von Staden and Oberholzer, 1977). Cohen and Reitz (1963) observed that TSS and TA of 'Valencia' orange and 'Ruby Red' grapefruit were similar to those for fruit grown on sour orange and superior to those on rough lemon or 'Rangpur' rootstocks. In some instances, TA is higher for fruit grown on trifoliata, a factor which may delay maturity. The fruit peel is smooth and thin, resulting in somewhat more splitting than with fruit on rough lemon-type rootstocks.

Citranges

Citranges are intergeneric hybrids of sweet orange and trifoliate orange. The original crosses were made in Florida by W.T. Swingle beginning in 1897 following the severe freezes of 1894–1895 to incorporate the freeze-hardiness of trifoliate orange into sweet orange. Several citranges have been tested as rootstocks, including 'Rusk', 'Morton', 'Savage', 'Benton', C-35, 'Carrizo' and 'Troyer'. The last two actually arose from the same cross between 'Washington' navel orange (seed parent) and P. trifoliata (pollen parent) made in 1909 (strictly speaking, these are citruvels, not citranges) (Savage and Gardner, 1965). Although seedlings of 'Troyer' and 'Carrizo' appear identical, some horticultural characteristics, like tolerance to burrowing nematode, differ, with 'Carrizo' being more tolerant. 'Troyer' is widely used in California and Spain, and 'Carrizo' has been a commonly used rootstock in Florida. Several other citranges have proved promising in rootstock trials worldwide but currently are not widely planted.

Scion cultivars budded on citranges produce moderately vigorous to vigorous trees, somewhat similar or larger than trees on sour orange, but generally smaller than those on vigorous rootstocks like rough lemon. 'Carrizo' and 'Troyer' citranges have become widely planted rootstocks for oranges and grapefruit for several reasons. Fruit are seedy and have a high incidence of nucellar embryony and thus are easily propagated as rootstocks (Table 4.2). In contrast, 'Morton' and 'Rusk' citranges are also highly nucellar but produce few seeds, a factor that limits the economic practicality of using these citranges as rootstocks (Hutchison, 1977). Trees on 'Carrizo' will grow moderately well on sandy and sandy-loam soils; however, they grow poorly on high pH soils (Wutscher, 1979). Citranges, like trifoliate orange and other trifoliate hybrids, produce a 'bench' (an overgrowth of the rootstock) at the bud union with most scion cultivars.

Wutscher and Dube (1977) found that yields of red grapefruit on 'Morton' and 'Troyer' citranges were similar to those of sour orange and 'Swingle' but superior to those of 'Milam' lemon when grown on soils with a pH ranging from 6.7 to 7.6. Similarly, sweet oranges grown on 'Troyer' citrange in South Africa (Von Staten and Oberhotzer, 1977) or 'Carrizo' in

Table 4.2. Seed per fruit and per cent of nucellar embryos for ten citrange rootstock cultivars.

Cultivar	Scientific name	Seed/fruit	Nucellar embryos (%)
Citranges	*C. sinensis* (L.) Osb.		
Carrizo	× *P. trifoliata*	23	100
Cunningham		4	94
Morton[1]		1	100
Rusk		5	96
Savage[1]		14	100
Troyer		20	98
Uvalde		9	100
Willits		3	90
L-44-4		2	100
L-44-7		2	94

Source: Hutchison (1977).

[1] The actual parentage is the reciprocal of the cross indicated.

Florida (Hutchison, 1977) also produced moderate to high yields.

The original reason for developing citranges was to produce a hybrid with edible fruit which was more freeze-hardy than sweet orange. However, scions budded on 'Carrizo' are generally of intermediate freeze-hardiness, depending on the time of a freeze (Yelenosky and Young, 1977). Freeze-hardiness is less than that of trees on sour orange, 'Cleopatra' mandarin or 'Swingle' citrumelo but generally better than on lemon-type rootstocks (Young, 1977). Trees on 'Carrizo' are not tolerant of early winter or late spring freezes in subtropical climates because they are slow to acclimate in the winter, and readily deacclimate early in the spring.

Scion cultivars budded on most citranges will be stunted by CEV; however, trees are not affected by CTV or xyloporosis. The extent of stunting varies with the strain of CEV present. 'Carrizo' citrange is tolerant to burrowing nematode (O'Bannon and Ford, 1977), i.e. nematode populations decline around the roots with time. Nevertheless, some 'Carrizo' seedlings, which are probably gametic, are burrowing nematode susceptible (D. Kaplan, unpublished) and therefore the source of rootstock material is important. 'Troyer' citrange is not tolerant of burrowing nematode. Neither citrange is tolerant of citrus nematode (O'Bannon *et al.*, 1977). 'Carrizo' is moderately susceptible to phytophthora foot rot particularly in young plantings. Blight susceptibility is intermediate between rough lemon and sour orange (Young *et al.*, 1982), although tree losses due to blight have been extreme in some coastal flatwoods areas of Florida (M. Cohen, unpublished), and as a replant for a tree lost to blight, trees on 'Carrizo' have been affected as early as four years of age (L.G. Albrigo, unpublished). Nevertheless, sweet oranges budded on 'Carrizo' are among the most profitable combinations over the long term in Florida (W.S. Castle, unpublished).

Citrumelos

Citrumelos are intergeneric hybrids of grapefruit and trifoliate orange. The original crosses were made in Florida by Swingle in 1907, but since then several citrumelos have been produced. Currently, 'Swingle' citrumelo is the most widely propagated rootstock in Florida and is gaining in popularity worldwide since its release in 1974 (Hutchison, 1974). There is conflicting information about the vigour of trees on 'Swingle'. 'Swingle' has a semi-dwarfing effect on sweet orange trees, while grapefruit trees on 'Swingle' are quite vigorous (Wutscher and Shull, 1975). Trees tend to be larger than those on sour orange or mandarin rootstocks, however. These differences may be due to the presence or absence of viruses, although 'Swingle' apparently is not adversely affected by CTV, CEV or xyloporosis. Lemon and lime cultivars do not yield well on Swingle rootstock.

Scion cultivars budded on 'Swingle' grow well on sandy and loamy soils, but grow poorly on clays, high pH soils, or in poorly drained areas (Wutscher, 1979). 'Swingle' has moderate salinity tolerance and is moderately drought tolerant (Hutchison, 1974). Wutscher and Shull (1975) conducted a 9-year study in Texas that showed yields of red grapefruit on 'Swingle' to be better than those on sour orange and rough lemon. In Florida yields of 'Valencia' oranges are lower on 'Swingle' than rough lemon or sour orange rootstocks; yields on a per unit area are quite high, suggesting that trees on 'Swingle' would be quite productive at high densities.

Scion cultivars on 'Swingle' are quite freeze-hardy, similar in tolerance to trees on sour orange and superior to those on rough lemon types or 'Carrizo' citrange rootstocks (Castle, 1987). This enhanced hardiness appears to be a major advantage of 'Swingle' over 'Carrizo' as a rootstock in chronically cold growing regions.

Trees on 'Swingle' are tolerant to CTV, CEV and xyloporosis (Hutchison, 1974), but trees infected with tatter leaf virus are stunted and may produce a bud union crease. Susceptibility to other viruses is unknown. 'Swingle' is not tolerant to burrowing nematode but is immune to citrus nematode (O'Bannon et al., 1977). 'Swingle' is very phytophthora-tolerant and blight tolerance is moderate to good, but more data are needed since most commercial orchards on 'Swingle' are still less than 15 years old. As a nursery seedling, 'Swingle' leaves are very susceptible to citrus bacterial spot, which prompted the burning of millions of nursery trees in Florida in the 1980s. This susceptibility, of course, does not carry over to scions budded on 'Swingle'.

Fruit size of sweet oranges and grapefruit are comparable with those produced on sour orange and 'Carrizo' citrange rootstocks. In addition, Wutscher and Shull (1975) and Wutscher and Dube (1977) found TSS and TA of red grapefruit on 'Swingle' to be similar to that on sour orange and to be higher than those on rough lemon-type rootstocks.

'Swingle' is potentially one of the better all-purpose rootstocks for grapefruit and sweet oranges. It is highly nucellar, producing about 10–15% zygotic seedlings, moderately seedy and very vigorous in the nursery. Its disease- and freeze-tolerance are very advantageous qualities in the field. 'Swingle' is less suitable than sour orange on clay or saline soils and does not have the burrowing nematode tolerance of 'Carrizo' citrange. The desirable characteristics of 'Swingle' have made it a widely planted rootstock in Florida on all but calcareous or high pH clay soils. However, some studies suggest that net income for sweet orange trees on 'Swingle' were less than those on more vigorous rootstocks over a 14-year period (W.S. Castle, unpublished).

Several other citrumelos have been evaluated in rootstock trials in Texas (Wutscher, 1977) and Florida (Hutchison, 1977). Some of these rootstocks produce scion characteristics different from those of trees on 'Swingle'. Further field testing, however, of these selections is necessary and 'Swingle' remains the major citrumelo in use as a rootstock.

Other Citrus Rootstocks

Several hundred other rootstocks have been evaluated worldwide but in general none has consistently surpassed the overall performance of the rootstocks discussed previously. In particular, a number of mandarins and mandarin-hybrids have been evaluated including 'Orlando' tangelo, 'Changsha', *Citrus depressa*, 'Sunki', 'Sun Chu Sha', etc. (Wutscher, 1977). Most do not have the consistent yields and fruit quality of currently used rootstocks. The 'Hongju' red tangerine, however, is widely used in central China as a rootstock for sweet oranges. It is well-adapted to the rocky, calcareous soils and cool growing conditions of this region. Recently, several rootstocks with characteristics similar to those of sour orange but with CTV tolerance have been tested in Florida with favourable results. These include *Citrus taiwanica*, *Citrus myrtifolia* and *Citrus obovoidea*.

'Sun Chu Sha' mandarin may have promise in some citrus regions. A 14-year study in Florida suggested a high survival rate for 'Valencia' oranges on this rootstock on the east coast where blight is a major concern (D.J. Hutchison, unpublished). 'Sun Chu Sha' also appears to be tolerant of CTV and *Phytophthora* spp. Tree vigour of 'Valencia' oranges on 'Sun Chu Sha' after 14 years was similar to that of sour orange and 'Cleopatra', but greater than that of 'Carrizo' citrange. Yields were similar to those on 'Carrizo', sour orange and rough lemon, but superior to those on 'Cleopatra' mandarin. In contrast, Wutscher and Dube (1977) found that grapefruit yields were lower on 'Sun Chu Sha' than on 'Swingle' citrumelo, sour orange or 'Troyer' citrange. Further testing of 'Sun Chu Sha' under a variety of growing conditions is needed, but the rootstock appears to have promise in areas having calcium soils or where blight is a major concern.

'Rangpur' × 'Troyer' is a promising hybrid being tested experimentally in several citrus-growing areas including Florida and California. Sweet orange trees budded on 'Rangpur' × 'Troyer' are 30–50% smaller than those on standard rootstocks, and yields are comparable with those on trifoliate orange (Castle, 1987). Trees budded on 'Rangpur' × 'Troyer' are semi-dwarfing and are suitable for high density plantings. Scions budded on 'Rangpur' × 'Troyer' are precocious and begin yielding fruit 2 years after planting. This hybrid has poor phytophthora tolerance and is sensitive to high salinity and susceptible to citrus blight. It is not affected by CTV, but it is susceptible to xyloporosis and CEV. Scions budded on it are fairly freeze-tolerant for oranges. Fruit quality is superior to that produced on lemon types and slightly inferior to that produced on sour orange, 'Carrizo' citrange, or 'Swingle' citrumelo root-stocks.

Grapefruit was a rootstock of interest in the 1950s and 1960s; although tree vigour was excellent, yields were lower than those on rough lemon but similar to those on sweet orange and 'Cleopatra' mandarin (Gardner and Horanic, 1961). In addition, fruit produced on grapefruit rootstock were small and had poor internal quality. Yields of rooted cuttings of red grapefruit were lowest of any rootstock studied (Wutscher and Dube, 1977). Therefore, grapefruit is not widely used as a rootstock worldwide.

Several citrus hybrids and related species have also been evaluated for use as rootstocks or interstocks for dwarfing of commercial citrus including species of *Clymenia, Hesperthusa, Citropsis, Microcitrus, Eremocitrus, Severinia, Fortunella* and *Swinglea* (Bitters *et al.*, 1977). However, none has proven commercially acceptable. The Chinese box orange (*Severinia buxifolia*) is well-adapted to alkaline soils of Texas, but yields were lower than on other commonly used rootstocks (Wutscher and Dube, 1977). Protoplast fusion hybrids of these relatives with *Citrus* are now being examined for use as rootstocks in Florida (see Recent Advances, p. 107).

Dwarfing Rootstocks

There has been limited success in developing rootstocks that reduce citrus tree size permitting the use of ultra-high density plantings like those used for apples or peaches. In regions such as the USA, Brazil and Mexico, sufficient arable land has been available for large, low density plantings, thus there was little incentive to use dwarfing rootstocks. In areas where land is at a premium, such as Japan, Spain and Italy, tree size is controlled by pruning. Similarly, hedging and topping is used to control tree size in other citrus-growing regions. However, recently interest in dwarfing rootstocks has increased as land and water become limiting.

Several methods have been developed to reduce citrus tree size including dwarfing rootstocks, inoculation of rootstocks with CEV, and use of interstocks. Dwarfing rootstocks include 'Rangpur' × 'Troyer', which was discussed in the previous section, 'Rubidoux' trifoliate, 'Rusk' citrange, 'Koethen' sweet orange × 'Rubidoux', and procimequat ([C. aurantifolia × Fortunella japonica] × F. hindsii). The tree size of 'Valencia' orange and 'Marsh' grapefruit on these rootstocks was smaller than that of trees on rough lemon, usually by less than 50%. However, in Florida and California sweet orange trees have been grown commercially on 'Rangpur' × 'Troyer' that are more than 50% smaller than a standard size tree. Fruit must be removed in the first 2 or 3 years to establish a tree canopy rather than a bush. 'Flying Dragon' rootstock (a trifoliate orange) causes extreme dwarfing and may be useful in ultra-high density plantings.

Inoculation of trifoliate and trifoliate hybrid rootstocks with budlines containing strains of CEV has also proved effective for dwarfing and is being used commercially in Israel and Australia (Bevington and Bacon, 1977). Extent of dwarfing was most pronounced for navel orange trees on trifoliata, intermediate for 'Troyer' and 'Carrizo' citranges, and least for trees on 'Rangpur'. However, even for the trifoliate rootstock canopy surface area was reduced by only 51%. There was a positive correlation between time of inoculation and final tree size, with those trees inoculated early being more dwarfed than those receiving inoculation as large trees. Recent studies indicate that the graft transmissible dwarfing agent used in Israel consists of at least five viroids rather than CEV alone.

Bitters et al. (1977) have tested several citrus relatives as interstocks for lemons with varying degrees of dwarfing ranging from 25 to 75%. While some of these relatives have promise as interstocks, this practice is not widely used to promote dwarfing of commercial citrus because of the cost and time involved to establish an interstock tree.

MORPHOLOGY AND ANATOMY OF CITRUS ROOTSTOCKS

The citrus primary root, the radicle, is the first plant part to emerge from the seed and produces a taproot which supports the tree. Lateral roots then develop, most of which are located in the upper soil layers. Fibrous, or feeder roots, develop from the laterals, producing a dense mass of fine roots with a high surface area which increases the capacity of the root for water and nutrient uptake. Fibrous roots often produce root hairs, although in some cases hairs are not clearly visible (Castle, 1978). Root hairs may improve nutrient uptake, particularly of immobile ions such as phosphorus.

Depth and density of the citrus root system varies with rootstock, environmental conditions (see Chapter 3), soil type, and irrigation and drainage practices. Fibrous root density of citrus growing in Florida varied from 0.5 to 1.3 g dm^{-3} in sandy loam soils to 1.9 g dm^{-3} in deep sands to 9.3 g dm^{-3} in shallow, poorly drained soils (Castle, 1978). The surface area of fibrous roots of 'Carrizo' citrange 13 months after planting was 3137 cm^2 with an overall length of 152 m (Bevington and Castle, 1982). Differences among rootstocks become apparent even in the nursery. Seedlings of C. volkameriana, C. macrophylla and 'Palestine' sweet lime had pronounced taproots and were most vigorous after 2 years in a field nursery in Florida. Rough lemon, sour orange and 'Cleopatra' mandarin seedlings were intermediate in vigour and P. trifoliata and 'Carrizo' seedlings had poorly developed, compact root systems (Castle and Youtsey, 1977).

The rootstock influence on rooting depth and distribution also occurs for mature budded trees in the field. 'Orlando' tangelo trees on rough lemon or 'Palestine' sweet lime rootstocks growing in well-drained, sandy soil had 50% of their fibrous roots at depths below 76 cm (Castle and Krezdorn, 1973). In contrast, 'Rusk' citrange and trifoliate rootstocks produced 60% of their fibrous roots shallower than 76 cm. Maximum rooting depth of trees on rough lemon rootstock approached 5 m, while 'Orlando' trees on trifoliate orange and 'Rusk' citrange rootstocks were the shallowest-rooted of the group. As expected, trees on the most vigorous rootstocks such as rough lemon produced the largest trees, whereas those on less vigorous rootstocks produced the smallest trees. Deeper rooted species such as rough lemon are typically more drought-tolerant than shallower rooted species like P. trifoliata because they access a greater soil volume.

Soil conditions, in particular soil structure, also affect rooting depth and distribution. In soils with high clay content, compacted soils, or those with impervious hardpans or high water tables, 75% of the roots may be located in the upper 15–45 cm of the soil surface (Castle, 1978). Studies from Cyprus suggest that feeder root growth is severely restricted in soils with a clay content greater than 80% and in South Africa citrus planting is not recommended in soils having greater than 50% clay content. Therefore, rootstock influence on rooting depth and distribution becomes less apparent as the potential volume of rooting becomes restricted by these factors.

PHYSIOLOGY OF CITRUS ROOTSTOCKS

Rootstocks differ not only morphologically but also physiologically with respect to water and nutrient uptake and salinity tolerance. Leaf water potentials (Ψ) of 'Orlando' tangelo trees on rough lemon and 'Palestine' sweet lime were less negative, indicating less water stress, than those on sour orange or 'Cleopatra' mandarin (Crocker et al., 1974). The enhanced water status of fruit for trees on rough lemon probably accounts for the dilution of TSS and

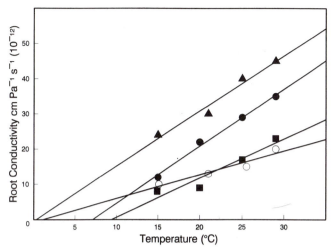

Fig. 4.1. The effect of temperature on the hydraulic conductivity of intact 12-month-old seedling root systems. Each point is the mean of four replicate plants of each rootstock. Linear regression lines have been fitted to data from each rootstock (R = 0.42 to 0.83). Symbols: ●, rough lemon; ■, sour orange; ▲, 'Carrizo'; ○, 'Cleopatra'. Source: Syvertsen (1981).

TA commonly observed for this rootstock (Albrigo, 1977). These differences in leaf Ψ occurred not only due to the greater rooting depth of the former two rootstocks but also due to greater root hydraulic conductivities which increased water uptake per unit of root mass. Syvertsen (1981) found that root hydraulic conductivity was greatest for 'Carrizo' citrange, intermediate for rough lemon, and lowest for 'Cleopatra' mandarin and sour orange (Fig. 4.1). The greater root conductivity of rough lemon over 'Carrizo' citrange may be partly responsible for the more vigorous growth of scions on this rootstock as well as for reduced freeze-hardiness (Wilcox and Davies, 1981). In general, there is an inverse relationship between vigour and freeze-hardiness of citrus (Young, 1977).

The general effects of citrus rootstocks on leaf mineral element content are summarized in Table 4.3 and represent years of research on this subject (Wutscher, 1989). It is difficult to make overall generalizations for the various rootstocks and elements, because in some instances a particular rootstock induces high levels of one element and low levels of another. However, it appears that rootstocks inducing vigorous growth in the scion such as rough lemon or 'Palestine' sweet lime (Castle and Krezdorn, 1973) generally have higher leaf levels of N and K than those low vigour inducing rootstocks like trifoliate orange. Syvertsen and Graham (1985) also observed that leaf N and P contents were positively correlated with root hydraulic conductivity of five citrus rootstocks. Leaf N and phosphorus levels and root conductivities were lowest for sour orange and highest for trifoliate orange seedlings. This trend

Table 4.3. Rootstock effects on mineral element levels in citrus leaves.

Element	Rootstocks inducing high levels	Rootstocks inducing low levels
N	Rough lemon Sweet orange 'Rusk' citrange Alemow 'Rangpur' lime	Sour orange Trifoliate orange 'Cleopatra' mandarin Grapefruit
P	Sweet orange Trifoliate orange Rough lemon Grapefruit *S. buxifolia* 'Swingle' citrumelo	Sour orange 'Cleopatra' mandarin 'Troyer' citrange 'Morton' citrange 'Milam' rough lemon
K	Grapefruit 'Sampson' tangelo *S. buxifolia* 'Swingle' citrumelo 'Milam' rough lemon	'Cleopatra' mandarin Rough lemon 'Rusk' citrange 'Morton' citrange 'Troyer' citrange
Ca	'Cleopatra' mandarin Rough lemon 'Troyer' citrange Sour orange	Sweet orange Grapefruit Alemow *S. buxifolia*
Mg	Trifoliate orange 'Carrizo' citrange 'Cleopatra' mandarin 'Sun Chu Sha'	Grapefruit Sour orange *S. buxifolia* 'Rangpur' lime
Na	'Rusk' citrange 'Yuzu' Rough lemon	Sour orange Sweet orange 'Morton' citrange
S	Rough lemon Grapefuit	'Cleopatra' mandarin Trifoliate orange
Fe	Rough lemon 'Yuzu' 'Rusk' citrange Sour orange	Grapefruit Trifoliate orange 'Taiwanica' orange 'Swingle' citrumelo
Mn	Rough lemon Alemow 'Yuzu' *S. buxifolia* 'Sunki' mandarin	Grapefruit Sour orange Sweet orange 'Swingle' citrumelo

Table 4.3. continued

Element	Rootstocks inducing high levels	Rootstocks inducing low levels
Zn	Grapefruit	Sour orange
	Rough lemon	Sweet orange
	S. buxifolia	'Carrizo' citrange
	'Cleopatra' mandarin	Trifoliate orange
Ca	'Rusk' citrange	Sour orange
	Sweet orange	Rough lemon
	S. buxifolia	'Troyer' citrange
	'Swingle' citrumelo	'Cleopatra' mandarin
Cl	'Troyer' citrange	'Sunki' mandarin
	'Carrizo' citrange	'Cleopatra' mandarin
	Trifoliate orange	'Milam' rough lemon
	Sweet lime	Sour orange
B	'Cleopatra' mandarin	S. buxifolia
	Sweet lime	Alemow
	Trifoliate orange	Sour orange
	Grapefruit	'Carrizo' citrange

Source: Wutscher (1989).

is not always consistent. Leaf N levels were not necessarily correlated with tree growth, as levels were among the highest for trees on 'English Small' trifoliate but trees were among the smallest (Castle and Krezdorn, 1973). Similarly, trifoliate seedlings accumulated higher levels of labelled ^{15}N than 'Swingle' citrumelo, yet growth was less (J.A. Lea-Cox, unpublished). Consequently, rootstocks that produce a small tree may have higher N on a percentage basis but lower total N than rootstocks producing large trees.

A rootstock's ability to exclude nutrients may be of more importance than its ability to accumulate them. Sour orange, 'Cleopatra' mandarin and 'Rangpur' (Kirkpatrick and Bitters, 1968), for example, are sodium and chloride excluders – qualities which make them tolerant of high salinity. *Severinia buxifolia*, which has limited potential as a rootstock, is an effective Cl and B excluder. In contrast, trifoliate orange, which is salinity sensitive, is a chloride accumulator. Leaf nutrient content, however, is not always a function of uptake rates as nutrients are also stored in the roots and may be translocated at different rates to the canopy. The physiological basis for differences in nutrient uptake by citrus rootstocks has not been thoroughly studied at the cellular level and the mechanisms for the observed differences in nutrient uptake are unknown.

MYCORRHIZAE

Some species of citrus rootstocks commonly become stunted in the nursery following methyl bromide fumigation. The cause of stunting was initially unknown but was later found to be due to an absence of vesicular arbuscular mycorrhizae (VAM) which increase uptake of immobile nutrients such as Zn, Cu and particularly P (Kleinschmidt and Gerdemann, 1972). Several mycorrhizal fungi are associated with citrus roots but most are *Glomus* species (Nemec, 1978). Endomycorrhizae are also commonly associated with citrus roots in the field, although the degree of colonization varies with tree age, P content of the soil and root, general soil nutrition and rootstock. Young trees generally have less extensive colonization than older ones due simply to differences in inoculation time. Mycorrhizae use metabolites which diffuse from the roots as an energy source. When root P content is high there is less leakage of metabolites and less germination of the fungus, with the inverse relationship occurring at low P content (Graham and Syvertsen, 1985). The energy expense by the plant in maintaining mycorrhizae is considerable. As much as 6–10% of labelled ^{14}C was translocated from sour orange seedlings to mycorrhizal fungi in only 2 h after incubation (Koch and Johnson, 1984). Mycorrhizal 'Carrizo' citrange and sour orange seedlings also have higher root conductivity than nonmycorrhizal seedlings (Graham and Syvertsen, 1984).

Rootstocks differ significantly in their dependence on mycorrhizae. In general, sour orange and 'Cleopatra' mandarin are most dependent, with sweet orange, rough lemon and 'Rangpur' intermediate, and 'Carrizo' citrange least dependent (Nemec, 1978). The degree of dependence is related to root density, hydraulic conductivity and ability to take up P. 'Carrizo' and *P. trifoliata* have denser fibrous root systems and greater hydraulic conductivities than sour orange or 'Cleopatra' mandarin. Therefore, these rootstocks are less dependent on the additional surface area provided by mycorrhizal fungi (Graham and Syvertsen, 1985), and thus the symbiotic relationship between the tree and fungus does not develop.

Citrus tree stunting in the nursery may be reduced by adding P or reintroducing mycorrhizae. At low levels of P, nonmycorrhizal seedlings of sour orange and 'Carrizo' citrange were stunted compared with mycorrhizal seedlings. However, as P levels were increased from zero to 560 kg ha^{-1}, plants became of similar size (Menge *et al.*, 1977). Mycorrhizae may be introduced by seed inoculation, banding in the planting row, or by layering in the soil (Ferguson and Menge, 1986). These practises are fairly effective under carefully controlled conditions. In practice, however, it is difficult and costly to culture the fungi because no *in vitro* cultural system exists. Generally, fungi must be cultured on an alternative host such as Sudan grass. In addition, colonies must be handled carefully to prevent desiccation in the field. Therefore, mycorrhizal fungi are not being widely used in commercial nurseries at present.

RECENT ADVANCES IN ROOTSTOCK DEVELOPMENT

Rootstock breeding and field testing are time-consuming processes, often requiring 20 years or more before new rootstocks can be released. The mechanics of traditional field breeding of citrus rootstocks is described in Chapter 2. In the 1980s and 1990s new methods of producing rootstocks are being developed based on biotechnology. These include protoplast fusion and genetic engineering (see Chapter 2 for details). Protoplast fusion allows for hybridization of species that might otherwise be sexually incompatible or which produce a high percentage of nucellar seedlings, making the development of new hybrids difficult. For example, Grosser and Gmitter (1990) have developed several hybrids which are being field tested in the 1990s. They have successfully developed hybrids of rough lemon and sour orange with the potential of developing a hybrid which has the CTV tolerance of rough lemon and the blight tolerance of sour orange. Traditional hybridization methods for these two species have proved largely unsuccessful due to the high percentage of nucellar embryos produced in the progeny. Several researchers are also developing methods of improving CTV tolerance in sour orange by inserting the gene for the coat protein of the virus into the sour orange genome (see Chapter 2). Although none of the new biotechnology techniques have yielded an improved rootstock to date, the potential for solving some of the major rootstock-related problems of citrus remains high.

5

PLANT HUSBANDRY

NURSERY OPERATIONS

Establishment of a reliable source of planting material is essential to the success of a citrus industry. Dissemination of diseased or genetically inferior trees can have catastrophic effects on the productivity of the citrus industry for years to come. Consequently, many of the major citrus-producing countries have stringent nursery regulations, and the establishment of a budwood registration programme is essential for the long-term success of developing as well as established citrus industries. Nursery operations differ from those in mature orchards because of higher tree densities and demands for resources. Nurserymen must be able to predict production trends, rootstock and cultivar demands and the availability of budwood, often 1 to 2 years in advance. Moreover, disease outbreaks like the nursery strain of citrus bacterial spot in Florida in the 1980s or severe freezes can virtually eliminate a nursery overnight. Traditionally, nursery trees were grown in the field, generally requiring 2 years from planting seed to reaching saleable size. Recently many greenhouse nurseries have been established that produce marketable trees in as little as 9–15 months (Moss, 1978; Castle and Ferguson, 1982). Various aspects of field and greenhouse nursery operations will be discussed in this chapter.

Site Selection

Site selection is extremely important to proper and successful establishment and operation of a citrus nursery. In selecting a field nursery site it is important to choose a warm location with adequate air and water drainage which preferably has not been previously planted to citrus. Use of virgin sites minimizes the danger of soil-borne disease problems such as phytophthora, pythium and nematodes. The location should also preferably have an adequate source of water and road access. Requirements for greenhouse nurseries are less stringent because the environment can be controlled and artificial media rather than the local soil are used. However, availability of water and power are very important.

Seed Selection

All nurseries must begin with a reliable, true-to-type seed source since nearly all citrus rootstocks are propagated by seed. It is important to purchase validated (true-to-type) seed from a reputable seed company or nursery, or the seed may be extracted by the nurseryman himself from rootstock mother trees maintained by the nursery. Seed should be visually inspected to ensure that it is true to type and properly formed and developed. The seed lot should contain uniform seeds, free of small underdeveloped material. Seed characteristics are quite distinctive for each cultivar, varying in shape, size and surface characteristics. For example, seeds of rough lemon and *Poncirus trifoliata* are small with a pointed micropylar end and a smooth seed coat. In contrast, grapefruit seeds are large and plump. Sour orange seeds have a wrinkled seed coat and a distinctly flattened micropylar end.

Seed numbers per litre and price vary considerably. For example, a litre contains about 5500 rough lemon or 'Cleopatra' mandarin seeds, 2500 sour orange or 'Swingle' citrumelo seeds, and only 2100 'Carrizo' citrange seeds. Although the standard units may vary worldwide, seed number per litre gives an indication of seed size. Price per litre also varies with availability and demand. For example, when 'Swingle' citrumelo seeds first became available and demand was high, price per litre approached US$500, whereas rough lemon seed sold for about US$35 per litre. The price per litre of 'Swingle' seed has decreased considerably since then.

Establishment of seed source trees and seed extraction are relatively specialized cultural practices. Seed source trees of the major rootstocks are generally grown from seed, although some may be budded on rootstocks. Fruit for seed extraction are collected in the fall after seeds have matured. Seeds extracted before mid-summer have a lower germination percentage than those harvested later (Fucik, 1978).

Seeds should be properly treated before planting. When the seed is removed from the fruit it is covered by a mucilaginous coating, which should be removed by washing or using an enzyme preparation (Barmore and Castle, 1979). As seeds are washed, off-types and underdeveloped seeds can be removed by flotation. After the mucilage is removed seeds generally receive two fungicide treatments, one a hot-water dip (51°C for 10 min) for phytophthora control, followed by treatment with a registered seed treatment fungicide to control other fungi.

Proper seed storage is essential for adequate germination and seed survival. Seeds should be thoroughly dried and placed in sealed plastic bags that permit gas exchange but limit desiccation. Seeds can be stored at 4–5°C for 6 months ('Mexican' lime) to 2 years ('Troyer' citrange and sour orange) without serious losses in germination percentage (Newcomb, 1977). Freezing must be avoided, however, as seeds may be damaged and germination reduced, especially at temperatures below –4°C.

Planting and Seedbed Preparation

Field nursery

The planted area should be disc-harrowed and levelled prior to planting. Some nurseries also treat the site with a preplant fumigant to kill soil-borne organisms and weed seeds. If this treatment is to be done the area should be disced and covered with plastic (if a volatile fumigant is used) after which the fumigant is applied. The site should be left fallow for 2–4 weeks prior to planting to reduce toxicity to newly set citrus seedlings. Fumigation also kills beneficial mycorrhizal fungi which aid in uptake of nutrients, especially phosphorus (see Chapter 4). Consequently, planting of some rootstocks, notably sour orange, into fumigated soils requires addition of extra phosphorus for adequate growth. Some companies also market mycorrhizal fungi which can be added to the site following fumigation; these fungi must be added under particular environmental conditions and may not be as effective as natural populations.

Seeds of the desired rootstock are generally planted in the spring in seedbeds after the soil temperature reaches 12.5°C or above. Presoaking seeds in water usually results in more uniform germination. Seeds are planted about 0.5 cm deep at 0.5–1.0 cm apart in the row, although spacing may vary depending on the vigour of the rootstock. Spacing between rows ranges from 15 to 30 cm. It is essential to maintain optimum soil moisture conditions during this time, without over-watering. Many nurserymen have overhead irrigation sprinklers that can also be used for freeze protection in some subtropical areas with pumping capacities of $0.2–0.6$ cm h^{-1}. Water should be free of phytophthora propagules and low in total dissolved solids to reduce risk of salt damage to the foliage.

Seed germination occurs 2–3 weeks after planting, depending on soil temperature and moisture. Optimum germination temperatures range from 25 to 30°C for most rootstock selections (see Chapter 3). This temperature range also provides optimum seedling growth rate. Percentage germination varies considerably with rootstock and environment but 75–80% is considered commercially acceptable (Castle and Ferguson, 1982).

Because seedlings have a very limited root system they are extremely susceptible to environmental stresses. The seedbed requires frequent irrigation over the entire growing season. Seedlings should receive frequent (once or twice a week) fertilizations with low analysis materials. Nitrogen, phosphorus and potassium should be applied at $1:1:1$ ratios in most citrus regions, although phosphorus rates may need to be increased in fumigated soils where mycorrhizal fungi have been eliminated. Fertilizer rates vary considerably between nurseries, but generally exceed 1000 kg of N ha^{-1} annum^{-1}.

Citrus seedlings are susceptible to damage by diseases and insects in the seedbed. Major soil-borne problems include foot rot (*Phytophthora* spp.) and damping-off (*Pythium* spp.). These pathogens cause extensive tree losses particularly in humid tropical growing regions. The foliage of lemon-type rootstocks is often affected by *Alternaria*, while scab (*Elsinoe fawcetti*) may be a problem for sour orange leaves. These problems can usually be controlled by fumigation, timely fungicide sprays and provision for adequate drainage. Insect problems include aphids, orange dogs, grasshoppers, whiteflies, and root weevils and leaf cutting ants in some areas. Damage by these pests can be prevented using standard control programmes.

Another important problem in the field seedbed is desiccation due to wind and sandblasting. This can be serious in open areas, especially during the spring when leaves have yet to develop a thick cuticle. Cover crops of grains or grasses stabilize the soil and provide windbreaks that help reduce damage to leaves and twigs.

Greenhouse nursery

Seed planting operations in the greenhouse vary from those in the field because growing conditions can be more carefully controlled. Seeds are exposed to ambient air temperatures prior to planting, using aerated water soaks for 8–36 h at 30°C (Castle and Ferguson, 1982). Seeds are then sown into one of a variety of tray types usually containing soilless media depending on individual grower preference and whether the plants are to be propagated entirely in the greenhouse or moved to the field nursery after budding. Most commonly used media include a mixture of peat moss or pine bark and perlite, vermiculite, styrofoam, or other inert material that provides adequate drainage. Dolomitic limestone is often added to the media to raise the pH of the peat moss to between 6 and 7. Some nurserymen add controlled-release fertilizers or micronutrients directly to the media. A number of commercial media are available, although many larger operators blend their own.

Seeds are commonly sown by hand into large rectangular styrofoam trays or into plastic tubular cells with pointed, tapered ends. Many greenhouse operations cull underdeveloped seeds at this stage to improve germination percentage.

Temperature, irrigation, fertilization and pest control can be carefully controlled in the greenhouse. Temperature is controlled using cooling pads with fans or louvred vents in the top of the house. Many greenhouses have

Table 5.1. Growth data of 'Hamlin'/'Cleopatra' citrus nursery trees as affected after 30 weeks by solution N levels.

Nitrogen level (mg l⁻¹)	Total scion growth			Root dry wt (g)	Shoot dry wt[1] (g)	Total dry wt (g)	Shoot : root dry wt ratio
	Length (cm)	Leaf area (cm²)	Dry wt (g)				
0	13.7	204.5	2.8	8.4	6.9	15.4	0.83
12.5	72.2	1395.7	20.4	15.0	26.3	41.3	1.79
25	80.0	1761.6	28.2	18.9	35.1	53.9	1.86
50	89.7	1930.1	30.5	19.7	37.2	56.9	1.92
100	96.3	2081.8	32.3	19.3	38.6	57.9	2.05
200	91.6	1832.5	27.6	14.4	33.1	47.5	2.29
Significance							
Treatment	***	***	***	***	***	***	***
Linear model[2]	0.55(Q)	0.56(Q)	0.58(Q)	0.37(Q)	0.52(Q)	0.48(Q)	0.54(Q)
Linear plateau	0.77	0.76	0.73	0.42	0.71	0.65	0.74

Source: Maust (1992).

*** Significant at $P = 0.001$.

[1] Includes rootstock trunk and all scion flushes.

[2] R^2 values are given under their respective models and (Q) indicates the highest order of quadratic fit. Each number represents the mean of three blocks with four replicates per block.

automated sprinkler systems that irrigate and fertilize once or twice a week. Liquid fertilizer application rates and ratios are similar to those applied in field nurseries, but are often based on concentration of N in the formulation. Many nurseries apply from 200 to 300 mg l^{-1} of N; however, current studies (Maust, 1992) suggest that these levels are too high for some rootstocks, recommending rates of 15–19 mg l^{-1} (Table 5.1). Most nurseries probably over-irrigate and over-fertilize which they justify based on the value of the trees under the erroneous assumption that trees will become saleable sooner.

Pest problems can be greatly reduced with proper greenhouse management, but they can be quite severe because of the optimum environmental conditions present in a greenhouse. Generally, soil-borne problems are less severe in the greenhouse, but pests such as whiteflies and spider mites can be very serious if not properly controlled.

Nursery Practices

Field operations

When the seedlings are large enough for transplanting (4–8 mm in diameter), during late summer or fall, they are transferred to the field nursery. At this time, large (bulls) and small (runts) trees as well as trees with damaged or distorted root systems or bench roots (curved roots) are culled in most citrus regions (Castle and Ferguson, 1982). Large and small trees probably represent zygotic rather than nucellar seedlings and if not culled may affect subsequent tree performance. Growth and yields of scions budded onto zygotic seedlings may vary from that of nucellar seedlings, although this is not always the case. This process is called lining-out and the trees are called liners in some citrus regions. Some nurseries apply a fungicide dip at this time. Trees are generally transplanted at 1–2 m between rows and 15–60 cm apart within the row, depending on the size of equipment used for cultural operations. Some nurseries plant liners in paired rows with about 30 cm between rows and 1–2 m between pairs. Tops may be pruned to a standard height of 45–60 cm, but again this practice varies among nurseries.

Greenhouse Operations

The greenhouse nurseryman has three options for lining-out (planting-out) rootstock seedlings: (i) seedlings are transferred to a larger container in the greenhouse; (ii) seedlings remain in the original container; or (iii) they are lined-out into a field nursery as described in the previous section. Seedlings are generally transferred to larger containers after 3–5 months when roots have filled the container and trunk diameter has reached 3–6 mm. Nursery-men usually cull seedlings, using only uniform, healthy material free of distorted or bench-rooted seedlings. Most nurseries cull 20–50% of their seedlings at this time. This culling allows the nurseryman to produce uniform plants which are in demand by growers. Moreover, uniform plants from the nursery will probably produce uniform trees in the field. In contrast Rabe (1991), in South Africa, found that nursery trees propagated from bench-rooted liners were as large and productive as non-bench-rooted trees after 7 years in the field. Therefore, there is some question whether such extensive culling is necessary. In regions where nursery trees are scarce growers are more likely to bud and plant small or off-type trees. Seedlings are generally grown for about 3–4 months before budding. Soilless media similar to those used for seedlings are also used in the containers.

As a second option, some nurserymen plant the seed directly into large containers and grow them in the same container through budding and until they are ready for field planting. This has an advantage of saving on labour and materials for repotting but reduces seedling density and makes culling more difficult. The third option is to transplant the seedlings to the field nursery as liners and to treat them as described in the previous section.

Propagation

Most citrus trees worldwide are budded using the T or inverted-T method. Budwood for the scion cultivar is collected when buds are not growing. Budding may be done shortly after budwood selection or in some areas budwood is stored at 4–5°C in moist peat moss in plastic bags if cambial growth has not yet begun in the rootstock. Cambial growth (bark slipping) is necessary to permit insertion of the bud into the rootstock. Alternatively, budding may be done immediately after budwood selection even if bark is not slipping using a chip-bud. This method has a much lower percentage of bud take than inverted-T budding and is much slower.

Commercial propagators bud 1000–5000 seedling d^{-1} depending on the closeness and condition of the plants. Many budders work in groups consisting of the budder and wrappers who follow directly behind the budder and wrap the buds. Often lower shoots are removed from rootstock stems prior to budding to make placement of the bud easier. Professional budders usually have greater than 95% take and are typically paid on a piece-work basis.

The T or inverted-T bud consists of making a vertical cut about 1–2 cm long followed by a horizontal cut either above (T bud) or below (inverted-T bud) the vertical cut. The inverted-T bud is preferred in high rainfall field nurseries since it serves as a rain shield for the open wound. The bud is then cut from the budstick and inserted in the cut. The bud is wrapped tightly with plastic tape to prevent desiccation and to allow the cambia of the scion and rootstock to unite. Buds generally 'take' 2–3 weeks after budding. Budding height varies from 5 to 80 cm above ground level depending on the growing region. In many lowland tropical areas, trees are budded high above ground level to permit maximum protection from soil-borne diseases such as phytophthora. Some studies, however, suggest that budding too high (>30 cm above ground level) will reduce growth of the tree (Bitters *et al.*, 1981) and requires more time to produce a mature nursery tree which also is more costly to the nurseryman.

The position of the bud on the budstick affects percentage bud take. Buds located in the apical portion of the shoot have a higher percentage take than those at the basal end (Halim *et al.*, 1988). This observation may be related to the physiological maturity of the bud, but bud location is usually not considered in the budding procedure. Moreover, buds that develop during the winter months grow more rapidly after budding than those from shoots developing during the spring although the reason for this is not known.

After buds have taken, the plastic tape is cut and the bud is 'forced'. There are three major methods of forcing: (i) the entire rootstock above the bud is removed (cutting); (ii) the rootstock above the bud is cut about half-way through from the bud side and the top portion is bent and tied (lopping); and (iii) the rootstock above the bud is bent over and tied without cutting (bending or looping). Each method has practical advantages and disadvantages. Cutting removes a large mass of leaves and stems, permitting trees to be spaced very closely, especially in the greenhouse nursery. Cutting also is less labour-intensive than lopping or bending, requiring only a single trip through the nursery. In contrast, bending and lopping require wider spacing and at least two trips through the nursery, one to tie the trees and a second to remove the top after the bud has begun vigorous growth.

The three forcing methods are used to promote optimum growth of the new bud. If the rootstock is left intact, the new bud grows poorly due to competition between it and the canopy of the existing rootstock. Bending and lopping decrease competition for nutrients and water between the rootstock and new bud, yet still permit transport of hormones and carbohydrates to the bud. Cutting off the entire rootstock canopy completely eliminates competition with the bud but also eliminates sources of these metabolites. Some studies suggest that the canopy area above the bud can be removed if leaves are retained on the rootstock below the bud (W.S. Castle, unpublished).

Studies from Texas (Rouse, 1988) and Florida (Williamson *et al.*, 1992) suggest that bud growth is greatest if the rootstock canopy is lopped or bent compared with cutting (Fig. 5.1). Translocation of labelled photosynthates from the rootstock leaves to the scion was greatest for bent compared with cut or lopped trees (Williamson *et al.*, 1992).

Citrus trees are grown from seed in some regions but this is rarely done on a commercial basis. 'Tahiti' lime trees are propagated by marcottage (air-layering) in Florida. A 2–4 cm bark patch is removed from the branch and the

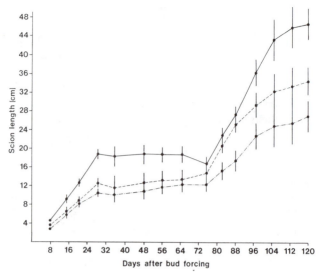

Fig. 5.1. Scion length of citrus nursery trees following bud forcing by cutting off, lopping or bending. Data points and bars represent means ± SE (*n* ranged from 5 to 22, depending on measuring date) (Williamson *et al.*, 1992). (—) Bending; (---) lopping; (-----) cutting.

exposed wood is covered with moist sphagnum moss, after which the moss is tightly wrapped with plastic. Callus tissue forms on the exposed surface followed by root formation. After roots have developed, the entire branch is removed and planted. Lime marcotts often produce fruit within a year of planting.

Cultural Practices in the Nursery

Fertilization, irrigation and pest control practices vary from nursery to nursery. Overhead sprinkler irrigation is commonly used in the field nursery, with application rates of about 2 cm per application, 1–2 times a week, depending on soil moisture levels. Fertilizer is applied once or twice per week in soluble forms (20N-20P-20K, 8N-0P-8K or 9N-3P-6K). A recent survey of nurseries in Florida indicated that fertilizer rates ranged from 1000 to 3000 kg of N ha^{-1} annum^{-1} (Castle and Rouse, 1990). However, plant density is also much greater in the nursery than in a mature orchard ranging from 32,000 to 63,000 trees ha^{-1} in the field to 170,000 trees in the greenhouse. Of interest, only 5–20% of the applied N could be recovered from the trees, suggesting considerable N losses due to over-fertilization. These figures contrast with 40–60% recovery of applied N under controlled greenhouse studies or in the field. Studies in Texas suggest that weekly application of soluble fertilizers produce greater shoot growth than applications every 2–4 weeks to containerized nursery trees (Rouse, 1982).

Major pest problems in the field nursery include orange dogs, spider mites, mealybugs, grasshoppers and root weevils. In the greenhouse, whiteflies, mealybugs and spider mites are major problems.

Sales and Distribution

Citrus trees are usually sold on a contract basis directly to growers or for cash to wholesale or retail nurseries for homeowner use. Contracts are often made before trees are produced to protect the nurseryman against economic losses resulting from unsold inventory. Contracts specify cultivar, rootstock, size limitations in some cases, date of delivery and price.

Containerized trees can be loaded directly on to trucks and need not be planted immediately. In contrast, trees to be sold barerooted, balled and burlapped or wrapped must be dug from the nursery, which removes a large percentage of the root system. Digging is done by hand or mechanically by pulling a specially modified blade under and around the tree. Barerooted trees are usually misted and covered with hay and plastic or canvas during shipping and may be heeled-in if there is to be a delay in planting. Some van delivery trucks have built-in misting systems to keep the trees moist during transport to the field and until distribution for planting (contract planting). Barerooted trees should be stored in the shade if possible before planting to prevent desiccation which may affect subsequent growth of the tree. Root desiccation before planting is a major cause of poor initial tree growth and even tree death (Grimm, 1956).

ESTABLISHING THE ORCHARD

Site Selection

Although citrus trees can be grown over a wide range of latitudes, climatic regimes and edaphic conditions (Chapter 3), proper site selection remains the key to successful commercial production. Several factors are important in site selection including geographic location and climate, soil characteristics, availability of water for irrigation, proximity to packing or processing facilities, availability of labour for cultural operations and harvesting, and costs associated with land and equipment purchases. Each of the above may be the limiting factor in site selection or more likely a combination of several factors may reduce the profitability of producing citrus in a given location.

Without question, low temperature is the critical factor dictating the range of citrus production worldwide. In subtropical regions where freezes occur on a regular basis it is essential to collect long-term, localized temperature information as well as regional climatic trends. For example, long-term data on freeze occurrences in Florida suggest that a severe freeze occurs on average once every 10.4 years. However, during the 1980s severe freezes occurred on average every 2–3 years. Moreover, regional data does not account for localized, microclimatic differences in temperature due to topography or proximity to oceans or lakes. It is common to observe significant differences in tree freeze-damage among closely situated orchards or those differing by only a metre in elevation in subtropical regions.

Growth and development of citrus trees is also influenced by extremely high temperatures (>50°C), although young, succulent foliage may be damaged at temperatures >40°C. This situation rarely arises except under specialized conditions such as in arid regions where leaf temperatures are typically 10°C higher than air temperatures. However, even under these conditions, leaf damage or distortion only occurs rarely except under unusual circumstances. Therefore, high temperature limitations on citrus production *per se* do not occur unless water is also unavailable, thus producing extreme water stress. Certainly, high temperature greatly increases transpiration, induces stomatal closure and decreases photosynthesis (Kriedemann and Barrs, 1981).

Commercial citrus cultivars can be and are grown over a wide range of edaphic conditions ranging from coarse impoverished sands (sand culture) to sandy loams to moderately heavy loamy clay soils or even muck soils. Nevertheless, the greatest productivity and tree growth occur in deep sandy to sandy loam soils, provided that temperature, light and water are not limiting factors. Citrus tree growth is reduced in poorly drained soils or where impervious clay layers or hardpans are present near the soil surface. Root and tree growth are also restricted in soils having greater than 50% clay content.

Other nonstructural aspects of the soil also affect citrus growth and should be considered in site selection. Citrus trees generally grow best at a soil pH between 5.5 and 7.0 due to improved availability of most important nutrients, although certainly several successful exceptions are notable worldwide. A considerable amount of citrus is grown at a soil pH between 7.5 and 8.5 with no major problems provided that the appropriate rootstock is used (see Chapter 4). In many of these areas, however, micronutrient deficiencies, particularly iron, are common. Citrus trees grown in low pH soils are subject to aluminium toxicity where high levels are present in the soil. In many instances soil pH may be adjusted by addition of limestone (liming) to raise the pH, or sulfur to lower the pH. However, many citrus-growing regions have highly calcareous or acidic soils whose pH is difficult to adjust. Therefore, rootstock selection becomes extremely important in these areas as well as for adapting the tree to various soil types (see Chapter 4). In general, soil pH alone is not a major limiting factor in worldwide citrus production.

Lack of adequate quantities of good quality irrigation water limits citrus productivity in growing regions like Brazil, China, Mexico and even in many tropical areas and is the major factor limiting expansion of citrus acreage in arid regions such as Australia and Israel. In many arid regions water quality is marginal for citrus growing due to high salinity levels (see p. 134). Moreover, even in humid subtropical regions water may be unavailable due to inaccessibility (mountain regions of China), lack of delivery systems (Brazil) or competition with urban areas for limited supplies of water (Florida and California).

Choosing a site with adequate drainage is equally important for successful citrus production. Citrus roots are killed or damaged by anaerobic conditions in the field (see Water Management, p. 125), and will not grow well where water has been stagnant for protracted periods. Areas with high annual rainfall and poorly drained, heavy soils like those in some of the American tropics are prone to drainage problems. Furthermore, poor drainage exacerbates problems with phytophthora foot and root rot, pythium and other soil-borne diseases. Adequate drainage is equally necessary in regions using flood irrigation to prevent water from standing in low spots and causing root death.

Many new citrus plantings are being sited in isolated areas due to problems with availability of affordable land and increased urbanization. Consequently, transportation to packing and processing facilities may be important in site selection. In fact, inadequate transportation capability is a limiting factor to increased production and marketing in many Third World countries, or in remote locations in other regions. In developed nations, although sufficient roads or railways may be available, escalating transportation costs may limit the profitability of some citrus sites.

A final significant factor associated with site selection is the economic feasibility of purchasing and developing a citrus orchard and the long-term economic prospects for the property. Potential citrus investors must consider the initial costs for land, irrigation systems (if necessary), tree planting, and interest rates, realizing that in many cases the break-even point ranges from 6 to 10 years from orchard establishment (Ford *et al.*, 1989). Moreover, the regional and worldwide supply and demand situation is also important. At present (1994) supply exceeds demand and prices are expected to be low at least until the year 2000. However, an orchard planted today will be subjected to prices 4–25 years or more in the future. The situation further varies depending where fruit is produced and for what market (fresh or processed). Changes in production in Florida or particularly in Brazil will have a significant effect on processed fruit prices but less impact on fresh fruit prices. For example, high production of oranges in Brazil and Florida in 1993 significantly decreased prices of processed orange juice worldwide. Production changes in Spain, Italy or Israel will affect fresh fruit prices in Europe but will have less impact on prices in the United States due to differences in marketing channels of the primary commodities produced in the Mediterranean. Tremendous expansion of citrus acreage and projected yield increases in

China, Brazil and Mexico, and some potential for increased production in Florida and Cuba have the potential to decrease long-term prices and must be considered as part of any site selection decisions.

Orchard Design and Planting Density

Selection of the proper orchard design and layout and planting density will have a significant impact on future yields, fruit quality, cultural operations and net returns. The overall objective of any planting design is to capture as much sunlight as possible while still allowing equipment movement through-out the orchard. Many different planting options exist depending on growing region and climate, topography, cultivar and rootstock and, to a lesser extent, intended use for the fruit, fresh or processed. For example, in lowland tropical regions plant density (a function of number of trees within and between rows) is generally low to moderate. Tree spacing may range from 7×7 (205 trees ha^{-1}) to as wide as 9×9 m (125 trees ha^{-1}) due to the excessive tree vigour in such locales, although higher density plantings also occur. Even at such wide spacing it is common for trees to form hedgerows in 5–6 years after planting. Tree spacing in commercial citrus orchards in subtropical regions varies widely but ranges from as wide as 8×8 (156 trees ha^{-1}) in Florida to as close as 3×3 m (1111 trees ha^{-1}) in Japan to 1.5×3 m (2222 trees ha^{-1}) in mountainous areas of central China, although such high density plantings are not commonplace. Therefore, topography, heat unit accumulation (see Chapter 3) and water availability affect decisions about planting density and orchard design.

In areas where large expanses of arable land are available (e.g. USA, Brazil and Mexico) most orchards are designed in square or rectangular configurations. The term 'rectangularity' is used to describe the relationship between tree spacings between and within rows. A symmetrical orchard design with even dimensions between and within rows has a rectangularity of one. In most citrus-growing regions orchard design is rectangular rather than square, usually resulting from decreasing within-row spacings relative to between-row spacing. The resultant increase in within-row density produces a hedgerow planting which has more trees ha^{-1} than a square configuration. Maintaining relatively wide between-row spacing allows for movement of equipment. Therefore, it is conceivable to have the same plant density with widely different spacings within and between rows, e.g. a 6×6 m and 3×12 m spacing each have 278 trees ha^{-1}, yet the rectangularities differ significantly, $1:1$ vs $1:4$. Considerable research has been conducted in several agronomic and tree crops, including citrus, to determine effects of varying rectangularity on yields and fruit quality. In California, varying within- and between-row spacings did not have a significant effect on yields of citrus provided that plant density did not change (Boswell et al., 1975). This fact, however, does not suggest that this would apply in a climate that induces faster tree growth or that net economic returns will not vary as spacings are varied because cultural operations such as spraying and harvesting may become more difficult and costly as trees grow into a hedgerow.

Orientation of tree rows within the planting becomes important as the orchard design shifts from square to rectangular. Rectangular plantings that will become hedgerows should be oriented north–south to permit maximum light interception, particularly for plantings in subtropical areas that are considerably north or south of the equator. It is advisable to maintain the tree height at no more than twice the distance between canopy widths, again to maximize light interception thus improving the effective fruit-bearing surface of the tree (Wheaton et al., 1978). For example, trees should be no taller than 4 m if the distance between the canopy perimeters between rows is 2 m. The distance between canopy perimeters is called the width of the drive middle. This distance is considerably less than the trunk-to-trunk distance. As trees become too tall, fruit is produced primarily in the upper compared with the lower or interior portions of the tree. Additionally, total soluble solids (TSS) levels and fruit size are low in interior and lower portions of shaded trees (Boswell et al., 1970). Citrus-growing regions such as Japan, central China, and many areas in the Mediterranean region are mountainous and lack large expanses of usable, readily accessible land. In these instances orchards are terraced and planted on the contour to decrease soil erosion. Cultural practices and harvesting are difficult and costly in terraced plantings; often only hand labour is practised in such locations. With terracing on hillsides, most plantings are made on the south-facing slope (in the northern hemisphere) again to maximize light interception.

High density plantings

The concept of high density citrus planting has been operational for many years; however, the definition of high density varies considerably worldwide depending on climatic conditions. Cary (1981) defines high density as more than 500 trees ha^{-1}. However, it may be more utilitarian to use low (<300), moderate (300–700), high (700–1500) or ultra-high (>1500 trees ha^{-1}) density categories, although these too are based on arbitrary divisions. Within each production region the overall tendency has been to increase plant density to optimize land use and most importantly increase early returns on investment by increasing yields during the early years of the orchard (Wheaton et al., 1978). There is a direct positive correlation between yields and number of trees ha^{-1} during the initial phases of orchard development (Phillips, 1974; Passos et al., 1977). Nevertheless, yields of ultra-high density plantings (2000 to 10,000 trees ha^{-1}) are only higher than those of low to moderate densities (300 to 700 trees ha^{-1}) for the first 5–8 years after planting (Wheaton et al., 1990).

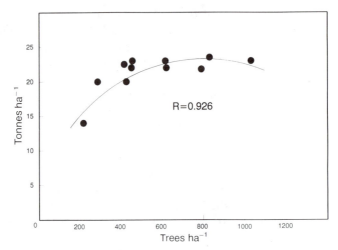

Fig. 5.2. Relation of 5-year average yield ha⁻¹ to planting density of 'Frost' nucellar navel orange trees. Source: Boswell *et al.* (1970).

Several studies worldwide suggest that moderate to high density plantings are more profitable than low density ones, even over a 15–20 year period, providing that they are properly managed (Boswell *et al.*, 1970; Cary, 1981; Koo and Muraro, 1982). A regular pruning or hedging and topping regime is necessary to control tree size and permits optimum light interception. It is also feasible that some trees will have to be removed within rows to improve light penetration into the canopy. In general, however, citrus growers are reluctant to remove any trees once production has begun.

Because of the tremendous variation worldwide in growing conditions no one plant density is optimum for every situation. However, long-term studies from the United States suggest that densities of 600–800 trees ha⁻¹ are economically optimum for 3–7-year-old navel orange trees under California (Mediterranean-type) conditions (Boswell *et al.*, 1970) (Fig. 5.2). This density range is intended to be a general guideline. Certainly, plant densities can typically be greater in cooler growing regions with less vigorous rootstock scion combinations, e.g. satsuma mandarin/*P. trifoliata* rootstock, such as those found in Japan (Tachibana and Nakai, 1989) and central China. Moreover, several tree density studies using *P. trifoliata* or related rootstocks dwarfed with citrus exocortis viroid (CEV) have achieved maximum yields at densities of 1000–5000 trees ha⁻¹, with optimum long-term yields at 1500 trees ha⁻¹ (Cary, 1981). In contrast, optimum plant densities are typically but not always lower in humid subtropical and lowland tropical areas with high heat unit accumulation and vigorous tree growth (see Chapter 3). Koo and Muraro (1982) found 300 trees ha⁻¹ to be the economically optimum density for 'Pineapple' orange trees over a 15-year period in Florida, although current studies suggest densities up to 500 trees ha^{-1} may be optimum (W.S. Castle, unpublished). Tree growth rate also differs among citrus-growing regions. For example, a 6-year-old sweet orange on 'Carrizo' rootstock growing in Costa Rica may be the same size as a 15-year-old tree of the same cultivar in Japan. Moderate to high density plantings should be used in tropical areas only if equipment is available for pruning or hedging and topping.

One factor associated with the effectiveness of high density plantings of some fruit trees like apple is the availability of dwarfing or semi-dwarfing rootstocks for controlling tree size. Although rootstocks affect tree size, few truly dwarfing rootstocks are available for citrus (see Chapter 4). Therefore, citrus tree size control is largely a matter of pruning and/or removal of trees as they grow together in a hedgerow, as in Japan. Such a system is more costly and difficult to manage than one based on a dwarfing or semi-dwarfing system. Lack of reliable dwarfing rootstocks has in the past been responsible for somewhat limited use of high density plantings in many citrus growing regions. However, introduction of dwarfing or semi-dwarfing rootstocks such as 'Flying Dragon' or 'Rangpur' × 'Troyer', or use of CEV to dwarf trees are two methods possibly useful in the future to restrict citrus tree size and are currently being tested in California, Australia, Israel and South Africa among others.

The primary objective of a high density citrus planting is to optimize light interception per unit land area, thus improving canopy-bearing volume and ultimately yields as rapidly as possible after planting. In traditional plantings, trees operate as individual units for many years until canopies grow together. For mature trees that have filled the space available to them, as within- and between-row spacings decrease where no hedgerow has been formed, canopy leaf area per tree increases but total canopy-bearing volume is nearly constant or decreases as the trees mature and grow together (Wheaton *et al.*, 1978). In contrast, in a hedgerow system where the within-row trees are allowed to grow into a continuous hedge, a decrease of between-row spacing increases canopy surface area and most importantly total bearing volume.

Wheaton *et al.* (1978) calculated a hypothetical change in bearing volume over a 30-year period for individual citrus trees compared with a hedgerow situation. Again, yield potential and bearing volume increase more rapidly for hedgerows compared with individual trees and as planting density increases. Therefore, the production efficiency (amount of fruit produced) is directly related to canopy bearing volume. Consequently, any high density system that increases canopy-bearing volume also has the potential to increase yields if other factors are not limiting, e.g. water or nutrients.

Potentially, high density plantings should make more efficient use of water and nutrients than lower density plantings; however, this supposition has not been documented experimentally. Water use (transpiration) should be a function of the total leaf surface area of the tree. Therefore, high density plantings would transpire more water on a per hectare basis than low density plantings during early development, although transpiration would become similar as orchards mature and trees fill their allotted volume. This has been demonstrated for high compared with standard density plantings in Florida. Nutrient requirements are a function of soil type, tree vigour and most importantly cropload in mature orchards. Again, during early orchard development high density plantings would have higher root densities ha^{-1} than lower density plantings because there are more trees ha^{-1}. Theoretically, the denser root system would intercept more of the applied nutrients, thus increasing the efficiency of nutrient uptake. This concept applies only where nutrients are broadcast over the entire orchard floor and not when using microirrigation systems or foliar sprays that deliver nutrients efficiently to a specific area of the orchard floor or tree. Moreover, as trees form a hedgerow, radiant heat loss during freezes is reduced, thus improving orchard temperatures and possibly lessening tree and fruit damage (Wheaton *et al.*, 1978). However, cold air may also collect in low spots causing more severe freeze-damage than in more open plantings.

While high density plantings may be potentially more efficient in use of pesticides, water and nutrients, some cultural practices are considerably more troublesome. Equipment must be adapted to operate in narrower drive middles and often spray materials do not penetrate sufficiently into dense hedgerows. Harvesting is especially difficult in hedgerows due to impaired movement of equipment and the reticence of pickers to harvest trees individually because of problems in moving ladders between trees.

Orchard Planting and Establishment

Proper planting and initial establishment of young citrus trees are essential for successful commercial development of an orchard. Although planting appears to be relatively straightforward, numerous trees have died because of a failure to follow such procedures. When planting barerooted trees it is important not to allow roots to desiccate. Grimm (1956) found that even short periods (1.25 h) of leaving trees unprotected in the field significantly reduced post-planting survival. Therefore, trees should be dug as close to planting time as possible and covered and moistened even if they are to remain in the field for less than 30 min. Similarly, trees in containers should not be allowed to dry out because the media will be difficult to rewet when the trees are planted in soil (Marler and Davies, 1987) due to differences in hydraulic conductivities between the soil and the media.

Rootstocks adapt the scion to various soil-related characteristics that might adversely effect a seedling tree (see Chapter 4). Sweet orange trees, for example, are extremely susceptible to phytophthora collar and root rots. Therefore, by budding well above soil level on a resistant rootstock the problem is alleviated. In many citrus-producing regions trees are planted too deeply, thus exposing the susceptible scion to foot rot and negating the value of the rootstock. Deep planting, a problem in most citrus-producing areas, can be simply resolved through proper instruction and supervision of planting crews.

Planting operations

Planting is fairly standard throughout the world. The orchard is laid out at a predetermined row and tree spacing. In some regions tree locations are determined visually by lining up rows with stakes at the end of the row. Many large-scale plantings are laid out using surveying equipment to ensure that rows will be straight, and spacings uniform. Irrigation lines are then laid out and emitters or sprinklers installed and tested (see Water Management, below) if in-place irrigation is to be used. In many areas, irrigation, particularly microirrigation, is necessary to ensure optimum production.

Tree planting consists of digging a hole that is the appropriate size for the tree, rather than pruning the roots or backfilling an overly large hole (Jackson and Tucker, 1992). Digging is done with an auger or by hand. Trees are then brought to the field either barerooted, burlapped, balled or in containers. Barerooted trees are often top-pruned in the nursery before digging because many roots are cut during digging. Burlapped or balled and container-grown trees are usually not pruned. Trees should be planted at the same depth as they were grown in the nursery and in most cases watered in to promote soil–root contact and remove air pockets that could cause root desiccation or the trees to settle bringing the scion to ground level.

There is some controversy as to whether field- or container-grown trees grow better after planting. Studies from Florida (Marler and Davies, 1987) suggest that field-grown, barerooted trees grow faster than container-grown ones for the first 2 years after planting. Moreover, research from Texas indicates that these differences in growth persisted for 10 years after planting; however, containerized trees although smaller were easier to manage and eventually had similar yields to field-grown trees (Maxwell and Rouse, 1984). Nevertheless, thousands of containerized trees have been planted with favourable results. The slower growth observed in these studies may have resulted from inadequate wetting of the media surrounding the roots or from the break-up of potbound root systems. Alternatively, container-grown trees are generally smaller in diameter than field-grown trees. Large citrus trees generally grow faster than small trees; however, these initial differences usually equalize within 5 years after planting.

WATER MANAGEMENT

Proper and successful water management is often necessary to achieve commercially acceptable yields of citrus fruits. Nevertheless, most citrus and related genera are water-conserving plants capable of withstanding long periods of drought when trees are mature. Therefore, it becomes necessary to separate the physiological adaptability of citrus to drought from the commercial necessity to obtain acceptable yields of high quality fruit.

Citrus trees are water-conserving plants due to a combination of anatomical and physiological factors that limit water movement through the plant (Kriedmann and Barrs, 1981). Root hydraulic conductivity (Lp) is inherently low, possibly due to the absence of well-developed root hairs and the presence of a fairly pronounced endodermis. As discussed previously, root hydraulic conductivity is positively correlated with root temperature and also varies with rootstock (see Fig. 4.1, p. 103). As root temperature increases from 15 to 30°C hydraulic conductivity increases (Syvertsen, 1981). Hydraulic conductivity is greater in vigorous rootstocks like 'Carrizo' citrange and rough lemon and lower in slower growing ones such as 'Cleopatra' mandarin.

Leaves of citrus trees are also adapted to conserve water. Stomata occur on the abaxial surface of the leaf. Stomatal control of water loss is quite poor in young developing leaves but becomes quite good as leaves mature (Syvertsen, 1982). Additionally, young leaves lack the structural rigidity of mature leaves and do not have as much epicuticular wax or cutin as mature leaves. Thus, young leaves are more likely to wilt during drought stress than mature ones. The waxy cuticle of mature leaves sometimes occludes stomata and is further responsible for limiting water loss from leaves. Citrus stomatal conductance oscillates throughout the day with a periodicity of 3 to 160 min depending on environmental conditions (Kriedmann and Barrs, 1981). This cycling is likely a function of microclimatic changes in CO_2 levels, temperature or vapour pressure deficit (VPD) (a function of temperature and humidity). Vapour pressure deficit has a particularly pronounced effect on citrus stomatal conductance even when factors like soil moisture are not limiting. During periods of high VPD (low relative humidity [RH], high temperature) stomatal conductance decreases, thus reducing water loss from the leaf.

Citrus water use efficiency (WUE), the ratio of the amount of CO_2 fixed per amount of water transpired, is quite low compared with many other C3 plants. This factor probably results from the relatively equal contribution of stomatal and residual conductances to total leaf conductance. Residual conductance, previously termed mesophyll conductance, refers to the movement of water and/or CO_2 through intercellular spaces and mesophyll cells. It also applies to movement of CO_2 into the chloroplast. Therefore, the WUE of citrus leaves is relatively constant. As water stress occurs, however, stomata begin to close restricting both CO_2 and water fluxes but residual conductances remain similar to stomatal conductance and WUE changes very little (Kriedmann and Barrs, 1981). In plants where stomatal and residual conductances have highly disproportionate values either CO_2 or water fluxes are preferentially altered and WUE either increases or decreases with water stress. Under extreme water stress, however, WUE decreases as net CO_2 assimilation approaches zero, since some transpiration still occurs.

Citrus fruit with their leathery rinds, low stomatal densities and high wax levels on the peel also contribute to overall water conservation by the tree. Classic studies from the 1930s suggest that fruits serve as reservoirs of water for the leaves during drought conditions at least for detached shoots. Years of field observations also suggest that trees with fruit withstand drought better than those without. Some researchers, however, have suggested that fruit serve this function only in detached shoots because detachment establishes abnormal water potential gradients between the leaves and fruit.

Water Stress

Water stress occurs when a lack of water causes the plant to function at suboptimal levels; it is manifested in several ways including a cessation of growth, leaf wilting or a decrease in stomatal conductance, net CO_2 assimilation or root conductivity. Water stress occurs when environmental conditions cause water uptake or transport within the tree to be insufficient to replace that lost via transpiration. Severe water stress inhibits vegetative and/or fruit growth, and causes leaf thickening and abscission.

Under most environmental conditions, however, citrus trees are exposed to moderate rather than severe water stress. Therefore, the key to successful water management is to minimize the amplitude and duration of the stress. Typically as temperature increases and RH decreases during the day, VPD increases (Marler and Davies, 1988). Stomatal conductance, transpiration and net CO_2 assimilation increase during early morning in response to these changes (Fig. 5.3). Transpiration increases largely as a function of increasing VPD creating tension in the xylem (water potential) and subsequent water stress within the plant as indicated by a decrease in water potential. Maximum stomatal conductance and net CO_2 assimilation of citrus trees generally occurs around mid-morning to midday with a characteristic afternoon decline in stomatal conductance and net CO_2 assimilation particularly at temperatures greater than 28–30°C. Temperatures above these values reduce net CO_2 assimilation primarily by reducing the activity of RuBisCo with a subsequent increase in internal CO_2 levels in the leaf. Increases in internal CO_2 levels then further decrease stomatal conductance thus limiting gas exchange.

Some degree of water stress occurs even under seemingly favourable

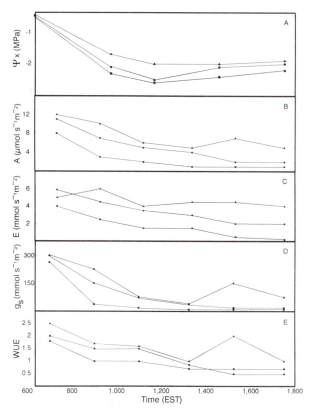

Fig. 5.3. Diurnal cycle of xylem water potential (Ψ), net CO_2 assimilation (A), transpiration (E), stomatal conductance (g_s) and water use efficiency (WUE) of young 'Hamlin' orange trees on 15 June 1987 as influenced by soil water depletion (SWD). ($\blacktriangle$) = high treatment at 20% SWD, ($\bullet$) = moderate treatment at 45% SWD, ($\blacksquare$) = low treatment at 60% SWD. Mean of eight measurements ± SE. Source: Marler and Davies (1988).

environmental conditions. As water is lost to the atmosphere via transpiration, the tree trunk, limbs and fruit lose water in a cyclic, diurnal pattern. A moderate level of water stress does not decrease yields or fruit quality provided water potentials do not reach critical levels that severely reduce net CO_2 assimilation or tree growth. Severe water stress occurs over time as soil moisture levels decrease and water becomes progressively less available to the plant (more negative water potential). Even so, the citrus tree itself can store large quantities of water in the wood and fruit which slows the decline in water potentials to critical values. Obviously, the amplitude and durations of these diurnal changes varies with temperature, VPD, light intensity, soil moisture levels and tree factors such as cultivar and rootstock. For example, Marler and Davies (1988) observed that the amplitude and duration of the water stress is usually greatest in high VPD and low soil water conditions (45 vs 20% soil water depletion) (Fig. 5.3). Stomatal closure serves to reduce the degree of water stress, but also reduces net CO_2 assimilation. In addition, water stress is often more severe in shallow-rooted rootstocks like 'Carrizo' citrange than in deep-rooted rough lemon types (Albrigo, 1977).

The objective of any successful water management programme is to provide sufficient soil moisture to replace that which is transpired during the day. However, it is important to realize that providing adequate soil moisture alone is not always sufficient to prevent water stress. In many cases, soil moisture may be at optimum levels, yet stomatal conductance and net CO_2 assimilation are low and young leaves may wilt due to lack of structural rigidity at reduced leaf turgor. This situation occurs when water loss occurs more rapidly than water transport from the roots to the leaves. This is why irrigation regimes based on soil moisture levels alone do not always accurately reflect tree water status levels.

Irrigation

The purpose of irrigation is to minimize the deleterious effects of water stress on growth, yields and fruit quality of citrus. Irrigation is necessary to achieve maximum growth and yields on a worldwide basis over a wide range of growing conditions (Kriedemann and Barrs, 1981). In general mature citrus trees require from 1000 (Koo, 1963) to 1563 mm (Van Bavel et al., 1967) of water a year to replace that lost by evapotranspiration (evaporation and transpiration, ET), although losses due to runoff and percolation may also be large (Hilgeman, 1977). The amount of irrigation varies with growing conditions. In arid or semiarid regions such as those in Israel (Bielorai et al., 1981), Arizona (Hilgeman, 1977) and South Africa (Bredell and Barnard, 1977) irrigation is essential to obtain optimum yields and tree growth and in some cases tree survival. Irrigation, however, also improves yields and tree growth in humid subtropical regions such as Florida (Koo, 1963) and São Paulo, Brazil. In fact, lack of widespread sufficient irrigation is a major factor limiting yields in Brazil. Furthermore, even in humid subtropical and tropical growing regions where sufficient rainfall is available for economic citrus production on a yearly basis, the seasonality and distribution of the rainfall may still adversely affect yields. Many tropical regions have pronounced wet–dry cycles and irrigation may improve yields even under these high rainfall conditions. Irrigation generally increases yields by increasing fruit size (Koo, 1963; Kriedemann and Barrs, 1981). However, irrigation also reduces physiological (June, November) fruit drop (Kriedemann and Barrs, 1981) and under extremely arid conditions improves flowering and fruit set.

Irrigation also affects fruit quality. In general, irrigation decreases fruit TSS and total acidity (TA) by a dilution effect and increases juice content.

Thus, excessive irrigation may decrease fruit quality (Hilgeman, 1977). Irrigation may also increase problems with oleocellosis (rupturing of oil glands in the peel) during harvest and enhance the incidence of zebra skin of mandarins and stylar-end breakdown of 'Tahiti' limes, both of which render fruit unmarketable.

Irrigation scheduling

Although the advantages of citrus irrigation have been clearly substantiated worldwide, the method of scheduling irrigation is still largely unresolved in most citrus-growing regions despite years of research on the subject. In theory, irrigation scheduling should be straightforward and based on the concept of replenishing daily water losses to evapotranspiration (ET), runoff and deep percolation (Rogers and Bartholic, 1976). Several methods of irrigation scheduling exist that are based on estimation or measurement of ET from an orchard. ET is a function of solar radiation, temperature, wind speed and RH. Temperature, solar radiation levels and wind speed are positively correlated with ET, and RH is negatively correlated with ET. Estimation of ET may be made from historical data for a particular location and time of year. ET may also be measured using a USDA Class A evaporation pan. The amount of water evaporated from the pan is proportional to the ET of the orchard but must be adjusted for plant factors such as root and stomatal conductances, leaf area, ground area covered by crop, etc. that regulate water losses from the system. Thus, plant ET is always less than potential ET and ET of citrus is not more than 80% of the potential ET. Correction factors (crop coefficients) relating pan evaporation values to citrus tree ET have been developed but still contain some measure of error and may vary with tree size.

Several irrigation studies have been conducted comparing citrus tree yields at various pan evaporation levels. Yager (1977) in Israel found no differences in yields of 'Valencia' orange trees at 35, 47.5 or 60% of pan evaporation. Swietlik (1992) made the interesting observation that pan evaporation coefficients varied with tree age from 0.75 in the first year after planting to 0.20 by the fourth year in the field. Thus pan evaporation coefficients are a function of the amount of ground surface area covered by the plant canopy and they should be adjusted yearly to compensate for these changes. Similarly, Smajstrla and Koo (1984) found no yield differences for 'Valencia' orange trees irrigated at 100, 50 or 25% of potential ET.

A quite common method of scheduling citrus irrigation is based on soil moisture levels – the idea again being that as moisture is lost due to ET, percolation or runoff, it can be replaced. This method is fairly straightforward and relies on soil tensiometers or other similar devices to measure soil moisture content. The soil moisture content at which to irrigate is usually determined by comparing the long-term relationship between yields or tree growth and soil moisture levels for a particular location and soil type.

Table 5.2. Effect of irrigation schedule on 'Valencia' orange growth, yields and fruit quality.[1]

No. of irrigations, amount of water applied (cm³ year⁻¹)	Trunk growth in 20 years (cm²)	Canopy growth in 17 years (m³)	Feeder roots at 0–183 cm in 17 years (g m⁻²)	Yield (20-year avg. wt tree⁻¹) (kg)	Fruit (20-year avg. wt fruit) (g)	TSS (20-year avg, Dec.) (%)
A. 15, 175	374a	107a	740a	125a	160a	10.6a
B. 10, 135	308b	76b	725a	115a	158a	11.0b
C. 15, 95	242c	66b	648b	88b	159a	11.4c
E. A Mar–July C Aug–Feb	244c	70b	697a	122a	140b	11.7d

Source: Hilgeman (1977).

[1]Numbers with different letters differ significantly, $P = 0.05$.

Therefore, a set of soil water depletion values can be derived for a particular soil type or location. Soil water depletion is a measure of how much water has been removed from a volume of soil relative to the amount of available water in the soil of the root zone. For example, a 20 year study from Arizona suggested that optimum citrus yields were obtained when soil moisture tension (as measured using soil tensiometers) remained between 60 and 70 kPa at 75 cm depth (Hilgeman, 1977) (Table 5.2). Vegetative growth increased with increasing irrigation but yields were comparable at the high and moderate levels. Other similar studies from Florida suggest maintaining soil moisture content at 33–66% of available soil moisture for optimum yields (Koo, 1963). Soil moisture depletion percentages of 30–45% (Marler and Davies, 1990) and tensions of 10–20 kPa (Smajstrla et al., 1985) are recommended to obtain optimum growth of young, nonbearing citrus trees. Young trees have more limited rooting and store less water than mature trees and thus require irrigation more regularly.

In theory, irrigation could most accurately be regulated using plant characteristics such as stomatal conductance, net CO_2 assimilation, leaf water potential or fruit or trunk growth. In practice, however, under field conditions some of these are difficult to measure and may not be accurate, reliable indicators of when to irrigate. For example, stomatal conductance and net CO_2 assimilation are affected by other factors besides plant water status such as atmospheric CO_2 levels, VPD or leaf temperature (Kaufmann, 1977). Similarly, leaf water potential is quite variable within the tree canopy during the day, although predawn water potential measurements may be used to assess tree water status. These methods, while useful for plant science researchers, are not readily adaptable for use by growers. Measurement of diurnal fruit or trunk shrinkage to ascertain citrus tree water status is more useful to growers but is quite tedious and not commonly used.

Therefore, due to problems and limitations to each of the above methods, many citrus growers irrigate based on historical patterns and the calendar system, or on a combination of soil-based and historical considerations. New, potentially more reliable and useful methods are being tested but as yet no single method has been adopted commercially on a worldwide basis.

Irrigation systems

Citrus trees have been irrigated for thousands of years, probably beginning with the use of catchment basins for rainwater in arid regions of the Middle East. Water was then diverted to citrus orchards using irrigation ditches. This flood or furrow irrigation method is still widely used in many citrus-growing regions worldwide. Berms (mounds of soil) are constructed to serve as borders along a row of trees after which water flows along the row until the entire land surface is covered. Water percolates through the soil mass to the root zone where it is taken up by the tree. In regions such as the poorly drained flatwoods areas of Florida or central China, citrus trees are grown on raised beds (ridges). In the limited areas of Florida still under flood irrigation, water is pumped or flows by gravity on to the orchard but then is pumped off after 48–72 h to minimize anaerobic root damage to the trees. Flood irrigation is also practised in Texas, northern Mexico and parts of central China.

Historically, many citrus orchards were irrigated using permanent overhead or undertree sprinklers or by travelling guns. In the early 1960s low volume (micro) irrigation systems were developed in Israel and South Africa to conserve water without compromising growth or yields. Several studies have clearly demonstrated that microirrigation uses less water than flood or sprinkler irrigation systems without compromising tree growth or yields (Roth et al., 1974; Koo, 1985; Swietlik, 1992). Roth et al. (1974) found yields of 'Campbell' 'Valencia' orange trees were comparable using trickle compared with flood irrigation, yet trickle irrigation used only 11% of the water applied with flood irrigation. Similarly, 'Ray Ruby' grapefruit growth and yields were comparable for trickle and flood irrigated trees, yet flooded trees received 24,410 kl ha^{-1}, while the trickle irrigated trees received only 1845 kl ha^{-1} over the 4 years of the study (Swietlik, 1992).

In arid or semiarid regions such as Israel, Australia, southern California or Arizona citrus tree roots are concentrated in areas where the irrigation is applied. In contrast, root distribution is more widespread in humid subtropical areas such as Florida where sufficient rainfall occurs to promote root growth. In these areas, it is important to irrigate as much of the root volume as possible to achieve optimum yields, particularly on deep, sandy soils (Koo, 1985). In general, more than 50% of the root system should be irrigated for mature trees in well-drained, sandy soils (Smajstrla and Koo, 1984). Similarly, Bielorai et al. (1981) in Israel found that yields of mature 'Shamouti' orange trees were greater when 90 or 70% of the root zone was irrigated compared with 35%.

Several microirrigation systems have been developed that deliver relatively small volumes of water at fairly frequent intervals, thus minimizing the large diurnal variations in soil and plant water status that commonly occur with furrow irrigation (Swietlik, 1992). Drip (trickle) irrigation emitters typically deliver 4–8 l h^{-1} and microsprinkler emitters from 20–80 l h^{-1}. With escalating costs and lack of availability of high quality water, microirrigation is becoming widely used throughout the world and will continue to be the method of choice in many citrus-growing regions.

Microirrigation systems provide an efficient method of providing water to the tree on a regular, consistent basis; however, they require more intensive management than other methods. Irrigation lines and emitters are subject to clogging by particulate matter, insects, spiders, or minerals such as calcium or magnesium which precipitate from the irrigation water. Various iron and sulfur reducing bacteria and algae also may cause plugging of emitters (Ford and Tucker, 1975). Therefore, water quality and adequate filtration and chlorination (to control algae and bacteria) are necessary to ensure proper operation of the system. Moreover, salts (Na, Cl) may accumulate in and around emitters especially in arid regions, in some cases causing root damage if they are not periodically leached from around the roots.

Salinity

Citrus trees are for the most part sensitive to salinity (Bielorai *et al.*, 1988; Alva and Syvertsen, 1991) so that water quality is an important consideration in any irrigation system. High salinity irrigation water may be a major limiting factor in citrus-producing regions such as Israel, southern California, Australia and coastal regions of Florida. In general, growth of citrus trees is impeded when saturated extract paste levels of the soil exceed 1.4 dS^{-1} (decisiemens). The situation is less straightforward when quantifying levels of total dissolved solids that cause leaf damage. However, salinity damage is intensified under low humidity conditions and where irrigation water with moderate levels of salinity (e.g. 1.3 dS^{-1}) is applied regularly.

In citrus-growing regions where high salinity water is common the trend has been to use microirrigation rather than furrow or overhead or sprinklers which apply water directly to the leaves or apply large quantities of salt. Choice of rootstocks such as 'Cleopatra' mandarin tolerant to salinity (Na and Cl) also decrease problems with high salinity water (see Chapter 4).

With increased competition between urban areas and citrus orchards for water, several regions use reclaimed wastewater for irrigation (Davies and Maurer, 1992). Secondary or tertiary treated wastewater provides an inexpensive and plentiful source of irrigation water and additionally may provide some necessary plant nutrients, thus reducing fertilizer requirements. Some reclaimed wastewater may contain high levels of Na, Cl or B, however.

Drainage and Flooding

Flooding (waterlogging) occurs when water displaces oxygen from the soil creating hypoxic (low oxygen) or anoxic (no oxygen) conditions. As oxygen is displaced by water or metabolized by microorganisms or roots, other compounds replace it as the terminal electron acceptor in respiration as the soil becomes reduced to <330 mV redox potential. Most soils suitable for growing citrus become reduced within a few days of flooding (Syvertsen *et al.*, 1983). For example, sulfides are reduced to hydrogen sulfide at a redox potential of about −150 mV by sulfur-fixing bacteria. Citrus roots are not adversely affected immediately by low soil oxygen *per se*, but are very sensitive to hydrogen sulfide levels in the soil. Consequently, the presence of hydrogen sulfide in the soil is a good indicator that the soil is reduced and suggests that citrus root damage is occurring.

The production of hydrogen sulfide and ultimately the extent of root damage is a function of soil temperature, organic matter content and microbial activity. Citrus roots are killed in as little as 3 d of waterlogging at relatively high soil temperatures (30–35°C) but may survive for months at lower temperatures (<15°C). Soils with high organic matter become reduced more rapidly than very sandy, low organic matter soils. Organic matter provides substrates for microbial reduction of the various compounds present in the soil. In contrast, citrus seedlings growing in sand culture survive months when flooded due to the absence of sufficient substrates for microbial production of hydrogen sulfide.

Citrus tree responses to flooding

Citrus trees growing in poorly drained sites are generally sparsely foliated, stunted and low yielding. In many instances trees do not die but remain marginally productive until the impediment to drainage is removed. In most cases trees are exposed to cyclic waterlogging and then drought stress related to wet–dry cycles, rather than to continuous flooding. Moreover, waterlogged soil conditions, besides debilitating the tree, are conducive to the proliferation of soil-borne fungi such as *Phytophthora parasitica* (root and foot rot) and *Pythium* spp. (damping-off). These organisms cause extensive tree death in nurseries, especially in tropical regions with high rainfall and poorly drained soils. It is advisable to grow trees on raised beds or in artificial media rather than soil in these locales. *Phytophthora* spp. in particular is a worldwide problem also causing extensive tree losses for young orchards (see Chapter 6). Symptoms for flooded compared with pathogen damaged trees are quite distinctive. *Pythium* causes seedlings to wilt and eventually die due to root death and girdling of the trunk. Foot or root rot symptoms include a pronounced chlorosis of the leaf midvein caused by root damage and girdling of the trunk. Lesions also appear on the trunk usually near the soil level (foot rot) or roots die and slough-off (root rot). Flood damage does not produce lesions.

Citrus trees respond physiologically to flooding long before morphological symptoms or yield reductions appear. Net CO_2 assimilation and transpiration decrease within 24 h of flooding and remain at reduced values as flooding persists (Phung and Knipling, 1976) (Fig. 5.4). Stomatal conductance also decreases as an adaptive mechanism to reduce transpiration because hydraulic conductivity of the roots decreases and thus water uptake is reduced (Syvertsen *et al.*, 1983). This survival mechanism, however, reduces net CO_2 assimilation which eventually translates to decreased shoot growth and yields.

As mentioned previously, soil temperature and texture have a significant effect on the severity and development of flooding damage in citrus. In addition, rootstocks vary in their flooding tolerance generally with rough lemon being more physiologically flood-tolerant than sour orange, *P. trifoliata*

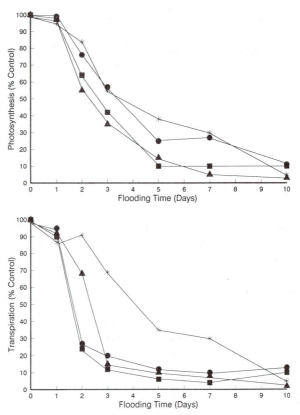

Fig. 5.4. Relative rates of photosynthesis (top) and transpiration (bottom) of four citrus rootstocks following flooding. Symbols: ●, rough lemon; ■, sour orange; ▲, 'Cleopatra' mandarin; ✳, *P. trifoliata*. Adapted from: Phung and Knipling (1976).

or 'Cleopatra' mandarin. Sour orange, and especially *P. trifoliata*, are more tolerant to foot rot than rough lemon; consequently under field conditions where the *Phytophthora* organism is quite prevalent, trees on rough lemon appear more sensitive to flooded conditions than those on sour orange or *P. trifoliata* (see Chapter 4).

FREEZE-HARDINESS AND FREEZE PROTECTION

Frost- and freeze-damage is of moderate concern in many of the largest citrus-producing areas worldwide, and is of major concern in some subtropical regions such as the United States, Japan, central China, northeastern Mexico and Argentina. Crop and occasionally tree losses have occurred through much of the Mediterranean region, as in Greece in the 1991–1992 season, but at irregular intervals and with far less severity than experienced in the United States. In contrast, several economically devastating freezes have occurred in the United States, most recently during the 1980s, causing the death of over 100,000 ha of citrus trees as well as billions of dollars in crop losses. Freeze-related crop losses rarely occur in Brazil, the largest producer of citrus worldwide, and never in tropical growing regions.

The definition of a frost as opposed to a freeze is somewhat subjective and controversial. A frost occurs under calm, clear conditions where extensive radiation heat losses occur. Frosts are common in Mediterranean-type climates with wide fluctuations in diurnal temperatures. A freeze connotes windy (advective) conditions which often cause more extensive damage than radiative freezes. However, since citrus tissues are damaged or killed only when ice forms, it seems most appropriate to use the terms 'freeze-hardiness' or 'freeze-damage' rather than 'frost-' or 'cold-damage' since plants do not sense cold (Rieger, 1989).

Freeze-hardiness

Commercial citrus and the most closely related genera are of subtropical and tropical origins and thus are intermediate in freeze tolerance between temperate zone species which acclimate to freeze conditions based on daylength and temperature cues and tropical species which have limited capacity to acclimate except through some supercooling. Citrus trees adapt to freeze conditions through a combination of freeze avoidance and freeze tolerance mechanisms (Yelenosky, 1985). Freeze avoidance involves the capacity of plant organs to supercool below 0°C without forming ice. The extent of supercooling varies with species, type of organ and extent of freeze acclimation. Flowers of commercially important *Citrus* species have the capacity to supercool to −4.3°C, fruit to −5°C, mature leaves to −7°C and stems to −8.9°C. However, young, succulent, nonacclimated expanding leaves supercool only to −2°C or less. Extent of supercooling also varies with species with leaves of *Citrus medica* (citron), *Citrus limon* (lemon) and *Citrus aurantifolia* (lime) supercooling the least (−3 to −5°C), and *Citrus unshiu* (satsuma mandarin) supercooling the most (−9.4°C) of the commercially important cultivars (Yelenosky, 1985). *Poncirus trifoliata*, when fully acclimated, has the capacity to supercool to −15°C, however, its primary use is as a rootstock and not as an edible cultivar.

During the 1980s the concept of ice-nucleating agents (INAs) and their potential role in citrus freeze-hardiness became popular. Ice nucleating agents are undesirable because they interrupt supercooling and serve as initiation points for ice formation and propagation through the plant. Although some research suggests that INAs may be important for citrus freeze-hardiness, lack of sufficient numbers of INAs under the field conditions tested and alternative mechanisms of freeze tolerance suggest that the role of these bacteria in citrus freeze-hardiness is less clearly defined than in herbaceous species (D.W. Buchanan, unpublished). Thus, attempts to reduce INAs and increase citrus freeze-hardiness in the field have been largely unsuccessful.

Citrus trees, unlike freeze-sensitive species, tolerate varying degrees of intercellular ice formation and consequently even if nucleation occurs cells will not necessarily be damaged or killed. Ice formation is manifested as dark-appearing areas (watersoaking) in the leaves usually emanating from the midvein toward the periphery of the leaf. Watersoaking in the field may occur at temperatures as high as $-3°C$ and may become widespread throughout the tree and orchard, indicating that nucleation occurs at numerous sites during a freeze. Nevertheless, the extent of watersoaking within the leaf does not necessarily correlate to the amount of leaf damage. It is common to observe fairly extensive watersoaking the night of a freeze without incurring leaf damage.

The degree of damage to the citrus tree is related to the amount of freeze acclimation and the duration and minimum temperature of the freeze. Degree of acclimation of citrus is a function of soil and tissue temperature and daylength. Citrus trees will not acclimate without light and acclimate more under long- than short-day conditions. The long-day effect is probably related to increased metabolite accumulation at long compared with short daylengths and not to phytochrome-mediated changes in metabolism as with temperate zone species (Yelenosky, 1985). Maximum freeze acclimation is attained under moderate daytime temperatures ($20–25°C$) and low night-time air and soil temperatures ($<12°C$) of 2 weeks or more. These conditions cause the plant to become quiescent or to have 'nonapparent' growth. This condition is not synonymous with true dormancy since placing citrus trees into favourable growing conditions (temperature $>12.5°C$) will cause a resumption of growth and a decrease in hardiness. Citrus trees deacclimate much more rapidly than they acclimate, sometimes within a few days at temperatures favourable for growth.

Numerous biochemical changes occur in citrus trees as they acclimate to low temperature, some of which may be correlative only, while others appear to directly affect hardiness. Sugar:starch ratios increase during acclimation as sugar levels increase dramatically (Yelenosky, 1985). Sugars may serve to lower freezing points within the plant or act as cryoprotectants for cell membranes. Trioses in particular are known to function in this way in other plants, although their role has not been clearly defined in citrus. Increases in proline in tree sap and changes in lipid and protein composition also occur during acclimation, but their functions related to freeze-hardiness are less well-understood.

The relative proportion of ice to water in the intercellular spaces and the rapidity of ice propagation through the citrus tree are also important in freeze tolerance. Differences in freeze tolerance among species may ultimately be a function of differences in the amount of frozen water tolerated at a given subfreezing temperature (Anderson *et al.*, 1983). Therefore, both minimum temperature during a freeze and duration of temperatures below the ice nucleation point are equally important to freeze survival. Ice nucleation occurs at more sites and ice propagation occurs over a longer period for long- vs short-duration freezes.

Methods of Freeze Protection

Passive methods

Without question the most successful method of freeze protection is site selection. Choice of areas where temperatures remain above −2°C are best to avoid fruit and tree losses due to freezes. It is very important to obtain long-term historical temperature records for a particular region and to assess the probability of freeze-damage. Avoidance of low spots that do not allow proper drainage of cold air (frost hollows) also reduces the risk of freeze-damage.

Other passive cold protection measures include wind breaks, clean cultivation and tree covers. Wind breaks reduce air mixing during advective freezes and thereby reduce heat losses from the orchard. Wind breaks are most effective if a heat source (like orchard heaters) is also present in the orchard. Wind breaks usually consist of closely planted rows of trees planted on the north or northwest border (south or southeast in the southern hemisphere) of the orchard. Man-made wind breaks (fences or shade cloths) also have occasionally been used, but are usually too costly to construct and maintain. Natural wind breaks (trees) may decrease citrus tree vigour and yields in the area nearest the break due to shading and competition for water and nutrients. Wind breaks should not be placed in low spots in the orchard where they can impede air drainage and allow freeze-damage in that area.

Clean cultivation is another method of passive cold protection. A clean, hard-packed surface intercepts and stores more solar radiation during the day and releases more heat at night than a surface covered with vegetation or a newly tilled area. Addition of water to the cleanly cultivated area prior to a freeze further improves heat accumulation during the day. Clean cultivation can be achieved using chemical or mechanical methods, provided sufficient time is given after tilling to allow the surface to become firmly packed.

Citrus trees have been grown for centuries by covering the trees (orangeries) to impede radiation losses and decrease freeze-damage (see Chapter 1). Orangeries have been used since Roman times and were common in northern Europe in the 15th and 16th centuries. The use of tree covers is not usually economical for large trees under commercial conditions, although some covered production still occurs around Sorrento, Italy. Moreover, the

Japanese grow citrus trees in greenhouse complexes with success. The greenhouse not only provides freeze protection, but also hastens fruit maturity by increasing temperatures, allowing early market fruit production.

Active methods

Mature trees survive moderate freezes of short duration better than young trees in most cases because of their greater canopy size and tree mass. The large canopy retards heat losses from the orchard and the greater mass requires long periods of low temperature to reach critical (damaging) levels. Moreover, young trees typically are more vigorous than healthy mature ones and thus are less quiescent and thus less freeze-hardy in many cases.

Orchard heating has been used successfully for freeze protection of mature groves for many years. The major advantage of orchard heaters is that, if properly placed and utilized, they provide effective freeze protection for trees and fruit, during both radiative frosts and advective freezes. However, heaters are costly to purchase, maintain and operate. Most growers can economically justify use of heaters only for high value fresh fruit or during very severe freezes where tree survival is jeopardized.

Wind machines provide another method of active freeze protection for mature orchards. They operate on the principle of taking advantage of the formation of a temperature inversion layer during radiative frosts. This situation is created when the soil surface cools faster than the air at 8–25 m above the surface. Wind machines, which consist of one or two large propellers powered by a gasoline or diesel engine, are used to mix the warmer upper air with the colder air in the orchard. Wind machines are most effective during moderate radiative frosts where only 1–1.5°C of temperature increase is required. They should not be used at wind speeds above 12 m s^{-1} (advective conditions) because air mixing has already occurred and also because high winds may detach the propellers creating a hazardous situation. Wind machines are not widely used because of high fuel and maintenance costs and damage by vandalism. Wind machines have been used along with heaters which provides better freeze protection than either method alone.

Various types of irrigation have been used for freeze protection of mature trees. Irrigation provides heat to the orchard as sensible heat and latent heat of fusion. In most cases the temperature of the irrigation water is 15–25°C when pumped from deep wells. As the water is applied to the orchard, heat is released to the air and trees. Another form of heat (the latent heat of fusion) is released as water turns to ice. Therefore effective freeze protection using water is dependent on a continuous, adequate supply of water. The major reasons for failure of irrigation to supply adequate freeze protection are insufficient quantities of water or inadequate coverage of the treated tree or area. These factors are particularly important during advective freezes where evaporative cooling occurs. Evaporative cooling removes 7.5 times as much

energy from an irrigated area than is provided by heat of fusion. Consequently, both minimum temperature and wind speed must be considered
when determining how much water is needed for freeze protection. Even
small increases in wind speed require relatively large increases in irrigation
application rates to provide adequate freeze protection. Severe freeze-damage
occurred in Florida in the 1960s and 1970s due to inadequate application
rates and poor coverage using overhead irrigation at $0.25 \text{ cm}^3 \text{ h}^{-1}$. Moreover,
there was extensive limb breakage due to ice accumulation. Therefore the use
of overhead irrigation is primarily limited to citrus nurseries where high
application rates ($1.0–1.5 \text{ cm}^3 \text{ h}^{-1}$) are adequate for freeze protection and
economically feasible because of the limited acreage that must be covered by
water.

Microsprinkler irrigation has become a popular method of irrigation in
many areas of the world. While microsprinklers at ground level are quite
effective for freeze protection of young trees, they are only marginally effective
for mature trees because they do not provide sufficient quantities of water to
the tree or irrigated area (Buchanan et al., 1982). Generally a temperature
increase of only 1–2°C is provided to the lower canopy by a typical ground
level microsprinkler system. This is because most systems apply inadequate
amounts of water. Microsprinkler irrigation generally provides no freeze
protection for the fruit and upper canopy, but may protect portions of the
lower trunk and leaves. However, microsprinkler irrigation has been effective
in protecting trees 2–3 m tall if emitters are elevated to about 1 m in the tree
canopy (Parsons et al., 1991). This method used on mature trees provides
freeze protection of the lower scaffolding, allowing rapid reestablishment of a
productive tree.

Flood irrigation is used for freeze protection in some areas of the world.
In most regions growers begin to flood the orchard the day before and during
the time when minimum temperatures are expected. Water is also pumped
onto the orchard the day before a freeze in Florida but is then removed within
48–72 h after the freeze to minimize root damage due to anaerobiosis.
Flooding an entire orchard provides 0.5–1.5°C temperature increase, mainly
from sensible heat. It is a fairly cost-effective and simple means of freeze
protection but is limited to areas using flood irrigation with access to a large
volume of water over a short time.

Care of Freeze-damaged Trees

Although thousands of hectares of trees have been killed by freezes worldwide,
historically freeze-damage generally affects only the fruit, leaves or wood to
varying degrees without killing the entire tree. This leaves the grower with the
major management decision of how to bring these trees back into production
as soon as possible. Tree age, freeze duration, cultivar and rootstock are

factors that determine how the orchard will be rehabilitated.

Assessing freeze-damage

The first step in managing freeze-damaged trees is to assess the extent of damage. As every freeze is different it is very difficult to make an immediate assessment of damage. Fruit damage is estimated by making cuts through the fruit at hourly intervals in the morning following a freeze. Some ice formation in the top 0.6 cm of the juice vesicles indicates mild damage, while solid ice formation in the centre signifies severe damage and loss of a portion of the crop. Generally, 4 h or more of temperatures of −2°C or below will cause some mature fruit damage. If extensive fruit damage has occurred, some fruit abscission occurs within 1–2 weeks following a freeze. High daytime temperatures following a freeze will, in particular, accelerate fruit drop and segment drying. Fruit should be harvested as soon as possible after a freeze and processed quickly to minimize decreases in juice content and yield losses due to fruit abscission. After ice in the fruit has melted, water is transpired through the peel rather than rehydrating the juice sacs, thus decreasing juice content. Some alternaria decay is likely to occur about 3–4 weeks after the freeze.

Leaf damage is difficult to assess during a freeze night. Watersoaked (dark green areas occur in part or all of the leaf) or curled leaves may or may not be significantly damaged. The morning following a freeze leaves may be rolled up and appear dry and dull green. These leaves will probably, but not always, abscise over the next week, again depending on temperature. Freeze-damaged leaves abscise between the petiole and the lamina (leaf blade) with the petiole dropping later. Leaf abscission is usually more extensive at the top of a tree than at the base following a radiative frost because temperatures are lower in this area due to direct exposure to the sky. It is not uncommon for temperatures of exposed leaves to be 1–2°C lower than sheltered air temperatures reported 1.5 m above ground level. Therefore caution should be exercised in interpreting minimum air temperature data relative to extent of freeze-damage. Within 1 week of a freeze the extent of leaf damage should be quite apparent. Trees can recover even from total defoliation and in some cases flowers and fruit will be produced in the next season, depending on when a freeze occurs, whether flower buds have already been initiated and the extent of wood damage.

The consequences of freeze-damage to twigs, stems and trunks is more difficult to assess than that to fruit or leaves. In general, small twigs will be damaged before larger limbs and trunks. Twig or limb dieback may not become visible for weeks after a freeze. It is common for large limbs to bud out in the spring following a freeze, only to die back in the summer or fall due to latent freeze-damage to cambial tissues. Another indication of wood damage is when leaves turn brown but do not abscise following a freeze. This indicates more severe freeze-damage than defoliation alone, and usually indicates severe limb damage.

Because freeze-damage to the wood is so difficult to assess, freeze-damaged trees should not be pruned until late spring or summer following a freeze. After the extent of freeze-damage has been assessed by evaluating the extent of cambial discoloration, pruning should be done to minimize problems resulting from melanose (a fungus which is harboured in dead wood). Pruning can be done by machine hedging and topping for minor damage or using chain or pneumatic saws when more selective, extensive hand-pruning is needed.

Cultural practices for freeze-damaged trees

Changes in cultural practices will probably have to be made depending on severity of the freeze-damage. It is important to assess freeze-damage accurately before altering cultural practices. In mild to moderate damage, partial or total defoliation with no wood damage, it is important to regrow the canopy as rapidly as possible. Trees should receive recommended fertilizer rates during the winter and spring and adequate but not excessive irrigation as new leaves develop. Most water loss is through the leaves and therefore it is unnecessary to apply heavy irrigation to defoliated trees. However, adequate soil moisture is important to promote uptake of nutrients and growth of new leaves. Weed control becomes a problem because the orchard floor receives more sunlight than a fully canopied orchard. Recommended rates of preemergence materials should be applied.

Cultural practices should be modified when severe leaf and wood damage have occurred. In this case, the size of the canopy and roots has been reduced and the tree requires less water and nutrients. For example, if canopy size is reduced by one-third, fertilizer, irrigation and herbicide rates should be reduced by that amount. Trees should receive more frequent applications of water and herbicide because of reduced tree size. There is also less demand for nutrients because a crop is not present but it is essential not to neglect these trees since new canopy development is essential. Buckhorned trees (those cut back to large scaffold limbs) may require hand-pruning to reshape the tree. Water sprouts that can lead to multiple trunks should be removed to lessen future management problems. Buckhorned trees may not require white-washing to reduce trunk temperatures if pruning is done in late spring allowing for regrowth of enough new foliage that will protect the trunk from high temperature damage. Trees buckhorned in the summer, however, require whitewashing to reduce trunk temperature and heat stress that may retard regrowth.

MINERAL NUTRITION

Nutrient Elements

Mature citrus trees require 12 elements besides carbon, oxygen and hydrogen which are readily abundant, to attain adequate growth and yields (Smith, 1966a). Elements required in large quantities (macronutrients) include N, P, K, Mg, Ca and S. Micronutrients include Mn, Cu, Zn, B, Fe and Mo. Citrus trees consist mainly of carbon compounds and water with nutrients comprising a small percentage of the total fresh weight (Chapman, 1968). Nevertheless, nutrients are essential to proper metabolic functioning of the tree and in ensuring consistent commercial production.

Abundance of macro- and micronutrients varies considerably with soil type and citrus-growing region worldwide. Some soils have inherently high levels of a particular nutrient or nutrients as well as a high cation exchange capacity. For example, P is rarely applied to mature citrus trees in the central ridge area of Florida due to high levels in the soil. Similarly, many high pH calcareous soils are high in Mg and Ca. Conversely, compounds such as K and particularly nitrate are commonly deficient due to rapid leaching from the soil and uptake by the tree. Therefore, application of N as urea, nitrate (NO_3^-), ammonia or a combination and K are essential parts of nearly all fertilizer programmes worldwide. Some studies suggest that as much as 40–50% of the applied N is not available to the tree due to leaching, denitrification (conversion of nitrate to N_2) and volatilization. Other essential elements are applied based on the inherent fertility of the soil and cropload.

Macronutrients

Nitrogen is a component of amino acids and proteins and is particularly important for proper growth and development of citrus trees. Adequate amounts of N are necessary to attain commercially acceptable growth and yields. Inorganic N is present in the soil solution primarily as N_2, NO_3^- or NH_4^+. Nitrate and ammonium are both taken up by citrus trees, although NH_4^+ absorption is greater at high pH and NO_3^- absorption at low pH (Kato, 1986). Several N fertilizers are available that provide these forms of N, none of which has been shown to produce superior yields compared with the others at least under Florida growing conditions (Leonard *et al.*, 1961). Nitrate is particularly mobile in the soil solution and may be leached from the root zone by excessive rainfall or irrigation. Moreover, both forms of N may be denitrified by bacteria to N_2O and N_2 (gases) which diffuse to the atmosphere. Nitrate is taken up actively (a process requiring energy) or in the transpiration stream and translocated to the canopy in the ionic form, while NH_4^+ is converted to amino acids, primarily glutamate, in the root after which it moves to the canopy in the transpiration stream (Kato, 1986).

Table 5.3. Standards for classification of the nutrient status of orange trees based on concentration of mineral elements in 4- to 7-month-old, spring-cycle leaves from non-fruiting terminals.

Element	Dry matter basis	Deficiency	Low range	Optimum range	High range	Excess
Boron (B)	mg l^{-1}	20	20–35	36–100	101–200	260
Calcium (Ca)	%	1.5	1.5–2.9	3.0–4.5	4.6–6.0	7.0
Chlorine (Cl)	%	–[1]	–[1]	<0.2	0.3–0.5	0.7
Copper (Cu)	mg l^{-1}	3.6	3.7–4.9	5–12	13–19	20
Iron (Fe)	mg l^{-1}	35	35–49	50–120	130–200	250?
Lithium (Li)	mg l^{-1}	–[1]	–	<1	1–5	12
Magnesium (Mg)	%	0.2	0.2–0.29	0.3–0.49	0.5–0.7	0.8
Manganese (Mn)	mg l^{-1}	18	18–24	25–49	50–500	1000
Molybdenum (Mo)	mg l^{-1}	0.05	0.06–0.09	0.1–1.0	2–50	100?
Nitrogen (N)	%	2.2	2.2–2.4	2.5–2.7	2.8–3.0	3.0
Phosphorus (P)	%	0.09	0.09–0.11	0.12–0.16	0.17–0.29	0.3
Potassium (K)	%	0.7	0.7–1.1	1.2–1.7	1.8–2.3	2.4
Sodium (Na)	%	–[2]	–	<0.16	0.17–0.24	0.25
Sulfur (S)	%	0.14	0.14–0.19	0.2–0.39	0.4–0.6	0.6
Zinc (Zn)	mg l^{-1}	18	18–24	25–49	50–200	200

Source: Smith (1966b).

[1]Indicates lack of information regarding value.

[2]These elements are not known to be essential for normal growth of citrus.

Nitrogen uptake and translocation are affected by several factors including soil temperature, root and tree vigour, and soil oxygen levels. For example, N uptake as well as reduction and assimilation were 10% lower for satsuma mandarins growing in the winter (low soil temperature) compared with the summer (high soil temperature) (Kato, 1986). Winter chlorosis or winter yellowing which occurs in many areas of the world may also be the result of reduced N uptake at low soil temperatures (<12°C). Chlorosis is most severe on mandarin hybrids such as 'Orlando' tangelo. It is also related to translocation of N metabolites at low temperatures. Kato *et al.* (1982) found that acropetal translocation of ^{15}N in the winter was only 0.1% of summer rates. These differences are likely due to reduced transpiration rates at lower temperatures as well as to direct temperature effects on active transport of N.

Tree vigour and rootstock also influence N uptake. Trees on rough lemon, a very vigorous rootstock, had greater N uptake rates than those on sour orange, 'Carrizo' citrange or 'Cleopatra' mandarin (see Chapter 4). Castle and Krezdorn (1973) also observed differences in leaf nutrient levels related to rootstock selection (see Chapter 4).

Optimum N levels for vegetative growth and yields are usually determined based on previous yields or through leaf analysis. Nitrogen-deficient leaves are distinctly light green to yellow in colour. The entire leaf becomes pale compared with the interveinal chlorosis which occurs with other nutrient deficiencies. Optimum levels of leaf N generally range from 2.5 to 2.7% for most cultivars. Ranges for leaf N for mature sweet oranges from several citrus regions are given in Table 5.3. These ranges are intended as general guidelines and may not apply to every situation and cultivar (Smith, 1966b). Leaf N concentration is often much higher for young, nonbearing trees than for mature trees, particularly just after transplanting from the nursery.

Phosphorus is essential for proper functioning of cell energy systems and as a structural component of cells; however, citrus trees require low levels of P. Phosphorus is present in the soil solution primarily as PO_4^{-3}, $H_2PO_4^{-2}$ or H_3PO_4 in a pH range of 6 to 7. Phosphorus is very immobile in the soil because it forms insoluble compounds with metals such as Al or Fe and tends to accumulate particularly in mature orchards. It is also leached and metabolized much more slowly than NO_3^- or K. Therefore, annual P application is not needed in many mature orchards. Phosphorus deficiency symptoms rarely occur on mature trees but include a pronounced reduction in bloom and poor production of small fruit. Leaf and soil analysis because of P immobility may be used to determine if P application is needed. Ranges of leaf P for sweet oranges in selected growing regions are given in Table 5.3 and are intended as general guidelines (Smith, 1966b).

Potassium is necessary for regulating ionic balances in the cell and for developing adequate fruit size and regulating peel thickness. Leaf K has little effect on vegetative growth of citrus within the range 0.35 to 2.0%. However, it has a significant effect on fruit quality. Low leaf K results in small fruit size and reduced peel thickness which predisposes the fruit to splitting, plugging (tearing of the peel around the calyx during harvesting) and creasing. Overapplication of K produces large, coarse fruit with thick peels. Potassium, like NO_3^-, is readily leached from the soil and is usually applied as a ratio of the N content either as muriate or sulfate of potash. For example, a $1:1$ N/K ratio might be used where a thick peel is desirable, while a $1:0.5$ ratio would be more desirable where fruit with a thin peel is more in demand. Potassium is naturally high in some soils and may not have to be applied. Ranges of K levels in sweet orange leaves in selected growing regions are given in Table 5.3 and are intended as general guidelines (Smith, 1966b). Potassium deficiency symptoms include production of small fruits with thin peels. Visual leaf symptoms are not as distinctive as those of N or Fe, therefore deficiencies are usually detected in leaf analyses.

Magnesium is essential to many enzyme reactions in the tree, but is not limiting in many orchards especially if dolomitic limestone (which contains Mg) is used to regulate pH. Magnesium deficiency decreases yields, tree vigour and freeze-hardiness; however, within acceptable levels Mg has little effect on vigour or yield. Deficiency symptoms include tip and margin interveinal chlorosis with a greener zone at the base of the blade. Sectional yellowing without distinct delineation between veins and interveinal areas of the older leaves as is observed with Fe or N deficiency is often present. Symptoms are more likely to occur in seedy cultivars or on calcareous soils that have received treatment with calcitic rather than dolomitic limestone. Magnesium deficiencies may also be discovered via leaf analysis (refer to Table 5.3 for general ranges of Mg) and are correctable using soil-applied MgO or $MgSO_4$, although leaf symptoms may not be corrected for over a year after soil application. Foliar sprays of Mg $(NO_3)_2$ (7% Mg) are also used to correct deficiency symptoms. Magnesium is most readily taken-up by the tree at pH values between 7 and 8.5. In most calcareous soils Ca competes with Mg for uptake sites on the root and limits uptake of Mg.

Calcium is the most prevalent element in a mature citrus tree comprising over 20% of the elemental content (Chapman, 1968). Calcium is important for enzyme functioning and is an essential component of cell wall structure and in transport of metabolites. In most citrus-growing regions Ca deficiency is rare since Ca is either abundant in the soil or added in lime to increase soil pH. Under controlled experimental conditions Ca deficiency decreases growth but effects on yields are not well-documented. Calcium deficiency symptoms are not distinctive but include a decrease in root and canopy vigour along with interveinal chlorosis similar to Fe or Mn deficiency symptoms. Calcium sources are most important to citrus culture for their role in increasing soil pH via liming.

Sulfur is an essential component of proteins and enzymes but is rarely deficient for citrus trees under commercial conditions (Smith, 1966b). Sulfur is usually provided as sulfate in many commercial formulations of fertilizer. Sulfur deficiency symptoms in leaves are very similar to those of N deficiency symptoms, but N deficiency is much more common.

Micronutrients

Micronutrients are necessary for proper enzyme functioning and, as the name implies, are present in small quantities in the tree. They are applied either to the soil or more commonly as foliar sprays. Soil-applied micronutrients become complexed in the soil and may be taken up slowly, requiring foliar application. Many growers apply micronutrients annually whether leaf deficiency symptoms are present or not. Long-term research from Florida suggests that regular applications of micronutrients are unnecessary and that growers should wait until leaf symptoms or leaf analysis dictate that micronutrients are needed (Wutscher and Obreza, 1987). The ranges of micronutrient concentrations for sweet oranges in selected growing regions are given in Table 5.3.

Manganese deficiency symptoms are common on young, expanding leaves appearing as yellow patches between veins. Such deficiencies do not require correction unless they become quite severe. Usually Mn is not limiting in acid soils but may become complexed and unavailable in high pH, calcareous soils. In this case Mn should be applied as a foliar spray.

Copper deficiency symptoms are rare but do occur in newly planted trees on virgin soils, in the nursery particularly for 'Swingle' citrumelo rootstock and occasionally in mature orchards. Symptoms include abnormally large leaves, limb dieback, gum pockets under the bark of young wood, brownish gumming on fruit, twigs and leaves, and multiple buds. Sufficient Cu is often applied in fungicidal sprays to prevent deficiencies, but Cu may be applied when necessary as a foliar spray along with other micronutrients or as a component of mixed fertilizer to the soil. Copper toxicity may occur where regular Cu sprays have been applied over the years. Copper becomes readily available to trees growing in acid soils and may become toxic when more than 100 kg ha^{-1} has accumulated in the top 30 cm of soil (Koo et al., 1984). Therefore, soil testing becomes a very important means of detecting excess Cu levels. Toxicity symptoms in the tree include reduced yields, production of small, dull green leaves, iron deficiency symptoms and stubby roots. If toxicity symptoms appear, soil pH should be maintained above 7 by applying lime to decrease the solubility of Cu in the soil.

Zinc deficiency symptoms are common worldwide and are readily distinguishable as distinct interveinal chlorotic regions. Leaves are generally smaller than normal and may occur in rosettes. Nevertheless, Zn levels alone rarely affect citrus growth or yields. Zinc deficiency is not readily corrected using soil applications, but is generally remedied by foliar application of ZnO or $ZnSO_4$ during the post-bloom period. Zinc deficiency symptoms commonly occur on trees affected by citrus blight, citrus variegated chlorosis, and greening (see Chapter 6) on leaves of freeze-damaged shoots and in sweet orange trees on 'Carrizo' citrange rootstock especially in high pH soils.

Boron level regulation in mature trees may be problematic since there is a narrow range between deficient, adequate and toxic levels (Koo *et al.*, 1984). Deficiency symptoms include corkiness of leaf veins and cork spots in the albedo of the fruit. Toxicity produces dull green leaves with brown surface pustules. Boron should be applied either in the fertilizer at 1/300 of the N rate or as a foliar spray. Boron should not be applied both by ground and as a foliar spray in the same year as toxicity may occur.

Molybdenum deficiency is manifested as interveinal yellow spots, usually in late summer. Deficiency symptoms are rare in most orchards, especially in high pH soils. Molybdenum deficiency is more severe on acid compared with neutral soils. Thus, deficiencies often can be corrected by liming to raise the pH or by foliar application of sodium molybdate.

Iron deficiency is a common problem and can have many causes. Chlorosis punctuated by a fine network of green veins is symptomatic of Fe deficiency. Severely Fe-deficient leaves may become light yellow to nearly white in colour. Iron deficiency, unlike other micronutrient problems, is not easily and readily corrected by foliar application of Fe. For soils with a pH greater than 7.0 (lime-induced chlorosis) FeDDHA iron chelate is applied to the soil or as a foliar application. On acid soils FeEDTA chelate is used. Chelated forms of Fe are not readily complexed into insoluble or unavailable forms. Deficiency symptoms are usually alleviated 6–8 weeks after application. Finely ground sources of chelated Fe, which are very expensive, should be mixed with an inert material and carefully applied to prevent burning of the fruit and leaves by drifting material.

Soil and Leaf Analysis

Soil analysis

Soil analysis can be used in conjunction with field observations, yield and leaf analysis in determining fertilizer and limiting rates or in correcting nutrient disorders. Soil analysis is most useful in monitoring levels of relatively immobile nutrients like Mg, Ca, Cu and P or for determining pH. Levels of these compounds in the soil are usually expressed in $kg\ ha^{-1}$ with low, medium and high ranges presented. Low levels of Mg and P can be corrected by applying these minerals as part of the fertilizer mixture. Soil N and K analyses are usually not representative of levels in the tree because of the mobile nature of these ions.

Proper soil sampling and extraction are necessary to ensure reliable results. Sampling is done using a soil auger or shovel by removing soil from a depth of 30 cm around the dripline of the tree. Usually samples from 16 locations within an area no greater than 8 ha are combined into a single composite sample (Koo *et al.*, 1984). This type of sampling is adequate if the soil in the orchard is uniform; however, composite sampling can be very misleading in variable soil types. The sample is then sent to a soil testing lab where it is dried, sieved and its pH determined. A sample of the remaining soil is extracted using a variety of acid mixtures, although the double acid (Mehlich) method is commonly used. The double acid method may cause excessive extraction of nutrients and does not correlate well with leaf analysis (Anderson and Albrigo, 1977). The extraction slurry is shaken for 5 min, filtered, and the extract analysed using atomic absorption (AA) spectrophotometry, inductive plasma spectrophotometry (ICAP) or colorimetry. It is important to be aware of the extraction technique used and how well its value correlates with citrus requirements for a particular soil type. Values are typically expressed in mg l^{-1} or kg ha^{-1}.

pH

Availability of some nutrients, notably Cu, Fe, P, K and Mg, is affected by soil pH. It is advisable to maintain pH at 5.5 to 7.0 if possible, although many orchards worldwide are productive at pH values considerably lower and higher than these. Liming of acid soils using calcitic (or preferably dolomitic limestone because it also contains Mg) will raise soil pH. Liming was a regular part of the production programme in the past but has become less important in recent years. In contrast, some orchards are planted in high pH, calcareous soils. In these soils Zn, Fe and Mn deficiency symptoms can become quite severe. Through a costly and relatively slow process pH can be decreased using soil additions like S, S-containing compounds and acid-forming fertilizers. The deficiency problems can be managed by applying chelated materials, but this is very expensive.

Leaf analysis

Leaf analysis is generally preferred to soil analysis for most nutrients. The leaves integrate uptake of nutrients over an extended period and the tree is the most accurate soil extraction system to indicate adequacy of each nutrient. A yearly programme of leaf analysis may be useful to citrus growers in assessing the overall nutritional health and fertilizer requirement of an orchard. As with soil analysis, proper leaf sampling is important. Samples of 100 leaves should be collected from at least 20 trees representative of the general orchard condition in no more than 8 ha units (Koo *et al.*, 1984). Fertilizer and cultural practices should be consistent over the sampled area.

   Several extensive studies have been done on effects of leaf age, location on the shoot and fruit load on nutrient levels (Smith, 1966b). Generally N, P and

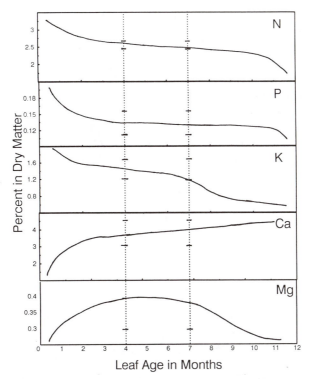

Fig. 5.5. Schematic diagrams of the change in concentrations of the macronutrient elements in orange leaves on non-fruit-bearing shoots as they increase in age. The dotted vertical lines indicate the age bracket of 4 to 7 months commonly used in taking indicator samples. The optimum ranges of concentration for leaves in this age bracket are indicated by cross marks for each element. The trends may vary somewhat under different conditions but these are fairly typical of the published results. Source: Smith (1966b).

K levels decrease from 1 to *c.* 4 months of age, while Mg and Ca levels increase (Fig. 5.5). Levels plateau for leaves between 4 and 7 months, with values decreasing for all elements as leaves age and nutrients are translocated to newer developing tissues. Similarly, leaf N and K levels increase with leaf age within a shoot, with newly developing leaves having lower levels than older ones. Conversely, levels of P, Mg and Ca remain fairly constant along the shoot. Leaf nutrient concentrations are also higher for non-growing compared with growing shoots, with the exception of P concentration which is relatively constant.

The presence of fruit on a shoot significantly decreases leaf nutrient concentrations of N, P, K, Mg and Ca in most instances (Smith, 1966b; Embleton *et al.*, 1963). This is expected since leaves compete for nutrients with developing fruits. This is evidenced by leaf N and K levels being inversely related to cropload (Smith and Rasmussen, 1961). For example, leaf N ranged from 1.97 to 2.31% in heavy crop years to 2.55 to 2.75% in years with light crops, with differences in K varying nearly twofold between heavy and light croploads.

Therefore, leaf samples are usually collected from nonfruiting, fully expanded, 4–7-month-old leaves in most citrus regions, with the exception of South Africa where samples are taken from fruiting shoots. If micronutrient or fungicidal Cu sprays have been applied it is very difficult to remove contamination, even with acid washing, and analysis will be inaccurate. Leaf samples should be washed three or four times with deionized water before air-drying and shipped in paper or plastic bags to an appropriate laboratory for analysis. Typical ranges of leaf analyses for sweet oranges in several citrus producing areas are given in Table 5.3 (Smith, 1966b). Values may differ slightly for grapefruit, lemons and other commercial cultivars (Embleton *et al.*, 1978). Leaf nutrient classifications include deficient, low, optimum, high and excess. These ranges refer to yield and fruit quality responses.

Fertilization

Citrus trees are grown on a wide range of soil types. Consequently, availability and inherent levels of nutrients may vary widely. In most areas, supplemental nutrients are needed to obtain commercially acceptable growth and yields. The type of nutrients required and the amounts are naturally a function of soil type, growing region and cropload. In impoverished soils with low cation exchange capacity (CEC) all the major macro- and micronutrients may be required to attain adequate growth and yields. In most soils suitable for citrus, however, primarily N and K, which are readily leachable, are supplementally needed. Growing region also affects the amount of supplemental nutrition required (Chapter 3). Regions with high rainfall and temperatures are more likely to lose nutrients from the soil due to leaching or volatilization. Cropload is also closely tied to fertilization programmes for mature trees since many nutrients are removed by harvesting the fruit. For example, a tonne of 'Valencia' oranges contains 1.31 kg N, 0.19 kg P and 1.8 kg K (Smith and Reuther, 1953). Therefore, an orchard producing 50 tonnes ha^{-1} would require 66, 9.5 and 90 kg ha^{-1} of N, P and K to replenish nutrients lost during harvest alone. This does not account for losses due to leaching or volatilization or utilization of nutrients for vegetative growth.

Fertilization of young, nonbearing trees

Fertilization practices for young, nonbearing trees differ from those required for mature, bearing trees because nutrients are not harvested with the fruit and the trees are much smaller. The objective in a young tree fertilization programme is to produce fruit as soon as possible by growing the tree as rapidly as possible. Productivity is highly correlated to canopy-bearing volume. Therefore, young trees receive high levels of nutrients, particularly N, to promote vigorous growth. In fact, N levels on a per treated hectare may approach 1000 kg because of the relatively small area that receives fertilizer. Several studies suggest that this rate is excessive.

A number of studies have been conducted on fertilization practices for young trees (Rasmussen and Smith, 1961; Marler et al., 1987; Willis et al., 1990, 1991; Swietlik, 1992). In general, young trees (nonbearing trees less than three years old) are fertilized more frequently than mature trees, 4–6 times per year vs 2–3 times, using smaller amounts of material for each application. This is due to the relatively limited root zone, particularly of a newly planted tree. Usually low analysis fertilizer is used to reduce the chance of salt damage due to over-fertilization. Fertilizer rates are based on tree age or size, soil type and growing conditions and may vary considerably among citrus regions.

Nitrogen is the most important element regulating growth of young citrus trees as during this time trees are growing very rapidly. It is not uncommon for tree trunk diameter to increase by 100 to 200% or more during the first and second growing seasons (Jackson and Davies, 1984), and canopy volume to increase tenfold in the humid subtropics. Other macro-elements, particularly P and K, are usually added to the fertilizer mix but have less impact on tree growth than N. In fact, studies from Florida indicate that leaf macronutrient levels from the nursery often are in the high to excess range. Similarly, leaf micronutrient levels are usually relatively high when the trees leave the nursery and thus supplemental sprays or soil applications are not needed unless indicated by leaf analyses or nutrient deficiency symptoms.

Traditionally, fertilizer has been banded or broadcast in a granular form; however, more recently liquid fertilizer is being applied through irrigation lines (fertigation) in many areas of the world. Some studies suggest that frequent application of low levels of liquid fertilizer improves tree growth over less frequent applications (Dasberg et al., 1988; Willis et al., 1991); the theory being that the concentration of nutrients is maintained at a constant level in the soil solution thus allowing continual uptake. Other studies, however, indicate that fertigation frequency or the use of granular compared with liquid fertilizer has no effect on tree growth, possibly because trees take up and store nitrogen as amino acids to be used for subsequent growth (Willis et al., 1991; Swietlik, 1992). It is likely that these regional differences in growth response to fertigation are due to edaphic, rootstock and environmental differences. For example, more frequent fertigation improved growth of 'Hamlin' orange trees on 'Carrizo' citrange rootstock but did not affect growth for trees on sour orange rootstock (Willis et al., 1991). Fertigation appears to have more effect on growth in areas such as Israel and Spain than in more humid regions like Florida and Texas. More effective and better-managed irrigation resulting from fertigation may also be involved in growth differences.

There has been considerable interest in using controlled-release (slow-release) fertilizers for young citrus trees. Controlled-release materials reduce the number of fertilizer applications needed per year and losses due to leaching (Marler *et al.*, 1987). Several products are available that have potential, especially in isolated reset (replant) situations. The major drawback to acceptance of controlled-release materials is the high cost per unit.

Fertilization of mature citrus trees

As stated previously, fertilization of mature, bearing citrus trees is necessary to replenish nutrients lost during harvest and leaching or volatilization, to maintain tree vigour and to obtain optimum yields (Smith, 1966a). Nitrogen has the greatest effect on tree growth and yields of all elements. For example, increasing the N rate for mature grapefruit trees in Florida initially produced a linear and significant increase in yields; however, at rates of above 200–250 kg ha^{-1} increasing N had a far smaller effect on yields. This law of the minimum, or the Mitscherlich effect, suggests that while N is necessary to obtain optimum yields, excessive N cannot be justified economically (Smith, 1966a). Nitrogen increases yields primarily by increasing fruit number rather than fruit size. Moreover, trees receiving optimum N are more densely foliated and produce more flowers than N-deficient trees.

Excessive N is not only economically unjustified but also may contaminate groundwater. Nitrate is highly water-soluble and moves rapidly through the soil profile into groundwater. High NO_3 levels in drinking water may be a health threat, especially to infants. Nitrate competes with oxygen in the bloodstream and in some instances causes 'blue baby syndrome'; however, this problem occurs very rarely and usually is not of major concern.

Several studies have been conducted worldwide to determine optimum rates and timing of fertilizer application, with, as expected, variable results depending on location, cultivar and intended use of the fruit (fresh or processed). Long-term studies by Smith (1969) in Florida suggest that about 900 g N tree^{-1} annually is necessary to achieve optimum yields for mature grapefruit, and that multiple compared with single applications has little effect on yields or fruit quality. Similarly, Mungomery *et al.* (1978) found that 900 g N tree^{-1} was the annual optimum for mature navel orange trees growing in Australia. They also observed that increasing N rates from 900 to 1350 g tree^{-1} did not increase yields, supporting Smith's previous findings (1966a). Twenty years of research from Brazil suggested that yields of mature 'Bahianinha' navel orange increased as N rates increased from 0 to 250 g of N tree^{-1} but no further increase was observed at 500 g tree^{-1} (Rodriquez and Moreira, 1969). These results were for non-irrigated trees, which may explain the lower optimum than in previous studies. They also observed that application of optimum levels of N and P together gave better results than either individually. DeVilliers (1969) conducted a 6 year study on 'Washington' navel orange in South Africa finding that about 1300 g N tree^{-1} annually produced optimum yields. Leaf mineral composition, growth and yields were compared over 9 years for 'Valencia' oranges in Arizona (Sharples and Hilgeman, 1969). Ammonium nitrate (granular) at nine annual rates from 0 to 3630 g of N tree^{-1} was compared with foliar urea sprays (230 g tree^{-1}). Little effect was observed on trunk growth related to rate or source; however, optimum yield occurred at about 600 g N tree^{-1}. Studies from Spain on 'Navelate' oranges again suggested optimum N rates in the 1000 g tree^{-1} range (Legaz *et al.*, 1981). Fruit weight tended to decrease with increasing N rate probably due to increased cropload. In this study, split application of fertilizer and fertigation were superior to single or granular applications. Studies from Israel also support this viewpoint, yet research from California (Jones and Embleton, 1967) and Florida (Smith, 1966a) suggest no yield advantage to split applications or use of fertigation for mature trees. Undoubtedly, climatic and edaphic differences between regions account for these conflicting results, although certainly optimum levels of N are similar worldwide. These rates might be reduced if leaching losses are minimized. Recent work (J. P. Syvertsen, unpublished) suggests that half to two-thirds of these previously determined optimum rates may be sufficient.

Citrus trees require relatively low levels of P to attain adequate growth and yields. The citrus fruit itself contains rather low levels of P which are removed during harvest and vegetative growth is not affected by increasing P levels provided that leaf levels are >0.08% dry matter (Smith, 1966a). Therefore, P deficiency in citrus is rare in most citrus-growing regions with the exception of some regions of South Africa. In fact, in the central Ridge area of Florida soil levels are naturally high (phosphate is mined there) or have been raised to high levels after years of fertilization. Thus, no P is added in the fertilizer. Sources of P for citrus fertilization include superphosphate or triple superphosphate.

Potassium fertilization is very important in many citrus-growing regions, particularly in fresh fruit producing areas where fruit size is an important concern. Nevertheless, leaf K levels vary considerably and studies suggest that vegetative growth is not significantly affected within a 0.35–2.0% range. Potassium is usually applied as potassium chloride, sulfate or nitrate.

Interactions of elements

There are several noteworthy interactions among the various nutrient elements. Nitrogen levels influence those of most other elements (Smith, 1966a). Nitrogen and P leaf levels are inversely related, with N levels having a pronounced effect on P levels. The N interaction with K has been widely studied. In general N and K levels are inversely related. However, the ratio

FACTORS INFLUENCED   PERCENT NITROGEN IN DRY ORANGE LEAVES

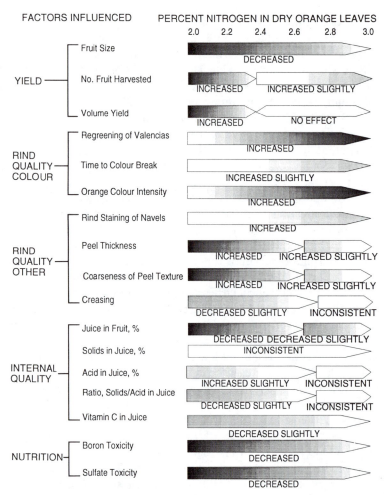

Fig. 5.6. Influences on yield, quality, and B and S nutrition resulting from changes in the percentage of N in 5- to 7-month-old, spring-cycle orange leaves. The greater the intensity of stippling, the greater the effect on the factor indicated. Source: Embleton *et al.* (1978).

between N and K has a significant effect on yields and fruit size (DuPlessis and Koen, 1988). Maximum yield for 'Valencia' orange trees in South Africa was obtained at a N:K ratio between 2.4 and 3.0. Both N and K had to be at levels greater than 2.1 and 0.8% dry weight, respectively. In contrast, optimum fruit size was attained at a ratio between 1.6 and 2.2 with N greater than 1.8% and K greater than 0.9%. Most importantly, greatest income was achieved at relatively low N levels. This situation occurs only for areas producing fresh fruit such as South Africa where fruit size and quality are directly related to income. Where citrus processing is more important than fresh fruit production, N levels should be in the moderate range to attain maximum yields. The ratios used in this study may differ from those in other citrus areas of the world due to climatic factors and since leaf nutrient levels in South Africa are determined for fruiting rather than nonfruiting shoots. Nitrogen and Mg levels in the leaf are positively correlated and synergistic, while conditions which produce high Ca leaf levels generally depress N levels. Potassium interacts with other elements besides N. Potassium and Ca displace one another in the soil and thus are mutually antagonistic. Thus concentrations in the leaf are inversely correlated. Similarly, NH_4^+ and K compete for sites in the soil. Potassium in addition is a strong antagonist of Mg, although Mg has only a moderate effect on K levels. There are several other interactions among elements but these have less effect than those mentioned above. For more detail on these interactions consult the review by Smith (1966a).

Nutrition and fruit quality of citrus

Nutrition and fertilization practices also influence fruit quality. Thus varying fertilizer rates and the ratio of elements is used commercially to change fruit quality depending on market demands. Nitrogen, P and K have the greatest influence on fruit quality, provided that other elements are not severely limiting (Embleton et al., 1978). The general influence of N, P and K on fruit quality also varies with species, with sweet oranges showing more response to N and P than lemons. Nevertheless, the general effects of changing leaf N levels from 2.0 to 3.0% dry weight for 'Valencia' and navel sweet oranges are given in Fig. 5.6 (Embleton et al., 1978). As leaf N levels increase fruit size, juice content and ascorbic acid levels decrease. Effects on TSS and TA are inconsistent. Increased N levels in addition increase peel thickness and coarseness of peel texture and decrease peel colour. Therefore, leaf N levels in the excessive range adversely affect fresh fruit quality. These same trends also occur for navel oranges. Increasing leaf P levels from 0.10 to 0.21% dry weight generally slightly decreases fruit size, TSS, TA, ascorbic acid, and peel thickness and coarseness. However, P levels have less effect on fruit quality than N or K. Potassium levels in the leaf have a pronounced effect on fruit quality as they increase from 0.3 to 1.9% dry weight. Fruit size and peel thickness and coarseness increase with increasing K levels and juice content decreases slightly. Increased K levels also reduce the amount of fruit creasing (see Chapter 7).

TREE SIZE CONTROL

Mature citrus trees can attain heights of more than 10 m and widths of near 10 m. In recent years the trend has been to decrease considerably tree spacing within rows, with a smaller decrease between rows. Such high density plantings have the advantage of earlier returns and higher yields, but also become hedgerows as little as 5 years after planting. As trees grow together, lower limbs become shaded and fruit production occurs primarily toward the outside and the top of the canopy of the tree. The interior of the tree becomes twiggy, limb dieback may occur and fruit size decreases. Yields and fruit quality decrease as shading continues. Trees that are allowed to grow unrestricted are difficult to harvest and spray. Consequently, as planting density has increased, interest in tree size control has also increased. In many citrus regions, trees have grown into canopy orchards and extensive and expensive rejuvenation pruning has been necessary to increase production and improve fruit quality. The most commonly used methods of size control include pruning and mechanical hedging and topping, although there is considerable interest in finding dwarfing or semi-dwarfing rootstocks or in the use of viroids such as exocortis to limit tree size (see Chapter 4).

Pruning

The two major types of pruning cuts are heading-back and thinning-out. Heading-back removes a portion of a branch and promotes lateral bud break, thus altering the morphology of the citrus tree. Heading-back cuts are usually made nonselectively using mechanical hedging and topping machines and are by far the most commonly used type of pruning for citrus trees in mechanized citrus-growing regions such as the United States, Brazil, etc. Thinning-out cuts selectively remove entire branches or limbs with a few cuts. Large portions of the tree can be removed, significantly changing light penetration into the canopy. Thinning-out cuts are generally used in conjunction with hand-pruning programmes. They also remove dead or unwarranted branches.

Selective pruning

Meticulous selective hand-pruning is used to control tree size and shape trees in citrus areas where labour is relatively plentiful and fruit are grown primarily for the fresh fruit market such as China, Japan and Spain. Hand-pruning is also common in these countries because most orchards are small and often family-owned. The system of pruning varies but generally involves hand-clipping small limbs and removal of dead or spindly wood from the interior of the tree, thus improving fruit distribution in the interior of the tree (Iwagaki, 1981). Tree canopies (skirts) are raised above ground level to minimize fruit damage from soil-borne problems, such as phytophthora. In some areas of Japan limbs are clipped immediately after fruit are harvested to moderate the next season's crop and to improve fruit size. In Spain, pruning and shaping usually begins when the tree is 2–3 years old. Scaffold limbs are selected around the trunk and other weak or poorly placed shoots are removed (Zaragoza and Alfonso, 1981). Interestingly, these researchers conducted a pruning study on 'Salustiana' sweet orange trees. They found that yields were the same whether trees were pruned annually or every 2, 3 or 6 years. Moreover, there was no effect on fruit size or quality. Thus, they questioned whether frequent, costly pruning was necessary, although it is commonly practised in Spain.

Selective pruning may be necessary to remove damaged or dead limbs, particularly after a severe freeze in subtropical areas. Pruning is done selectively using hand or pneumatic pruners or loppers, or chainsaws. Hand-pruning involves heading-back and thinning-out cuts of moderate to large limbs that cannot be pruned using mechanical hedging equipment. Selective pruning is also used to skeletonize (removal of a large portion of the canopy) or buckhorn (cutting back to major scaffold limbs) trees when they are to be top-worked or for rejuvenation of mature, nonproductive trees. Rejuvenation pruning is used only where trees have grown into a densely foliated canopy orchard and yields have become severely limited, but where tree structure is basically sound. The entire canopy is removed back to four to six major limbs to promote the production of vigorous new growth from latent buds. Rejuvenation eliminates production for about 2 years after pruning, after which yield will gradually increase. However, on a long-term basis the orchard will be more productive than if left unpruned. Pruning of this type is costly and produces large amounts of brush that must be removed from the orchard. Furthermore, severe pruning exposes the orchard floor to light and promotes growth of weeds.

Various types of pruning paints have been applied to pruned limbs over the years with varying levels of success. During the 1950s and 1960s the black asphalt-based paints were used but were found to result in excessive heat buildup on the wound area and to delay healing. White latex-based paints were developed to reduce limb temperature and promote healing and have been used by citrus growers for many years. Years of scientific study with forest trees, however, indicate that pruning paints do not improve wound healing and, in fact, may delay it by preventing natural wound periderm formation. Certainly, current hedging and topping practices produce millions of freshly cut limbs, most of which heal without painting the cut surfaces.

Hedging and topping

Hedging may be required on a regular basis by the fifth or sixth year after planting, depending on location, initial tree spacing and rootstock/scion combination. Hedging is typically accomplished using large circular saws vertically mounted to an adjustable mechanical arm. The cutting angle of the arm may be adjusted from 90° for routine hedging to 65–80° from horizontal for a combination of hedging and topping commonly called 'Christmas tree' topping. Hedging may be done any time during the year but is most effective if just before the spring flush, or after the threat of severe freezes has passed in subtropical areas (Phillips, 1980; Zaragoza and Alfonso, 1981). Hedging during this dormant period allows improved light penetration into the canopy during the spring growth flush, reduces canopy volume and promotes growth of lateral buds. Hedging at this time will not remove flowers (although flower buds may be removed) or fruit for most cultivars, with the exception of 'Valencia' oranges (which require greater heat unit accumulation for fruit maturation than other citrus cultivars) and occasionally grapefruit. Nevertheless, it has been shown that moderate, consistently timed hedging, even of those cultivars, will not significantly decrease yields. Fucik (1977) found hedging and topping of 'Ruby Red' grapefruit trees in Texas decreased yields in the following year compared with unpruned trees. Nevertheless, average yields over a 7 year period were similar and hedging and topping improved fruit quality and increased fruit numbers in the inside and bottom of the canopy. Similarly, Zaragoza and Alfonso (1981) compared manual pruning and hedging and topping with no pruning for navel orange trees in Spain. All pruning methods decreased yields compared with unpruned trees in the pruning year; however, yields were similar for all treatments in the next season when no pruning occurred. Most importantly, the average annual yield tree^{-1} was the same over a 6 year period for all treatments. Hedging of 20-year-old 'Valencia' orange trees in Cuba increased fruit size and yields over unpruned trees during all 4 years of the study (Borrel and Diaz, 1981). Hedging also decreased external blemishes on the fruit by allowing improved penetration of spray materials and increased air movement through the tree. An elaborate hedging and topping experiment was conducted over a 3 year period on lemon, mandarin and sweet orange trees. The treatments included: (i) hedging one side of the tree the first year, the other side in year 2 and topping in year 3; (ii) hedging both sides in year 1 and topping in year 2; or (iii) no hedging or topping. In general, moderate hedging and topping did not reduce yields of any of the cultivars but did improve fruit quality.

Several different patterns of hedging are used, none of which has proved to be superior to the others. Some growers hedge only one side of the tree in a given year, while others hedge both sides along the row in the same year. In the first case hedging costs are less on a yearly basis, but fewer advantages of hedging, viz., improvement of regrowth and increased light penetration into the canopy, are realized. In the second instance, tree width is controlled in both directions further improving light penetration into the canopy. In areas where orchards are not bedded (ridged), trees may also be cross-hedged to limit size in the row. Most effective hedging programmes will remove only the outer 15–20 cm of the canopy, thereby cutting limbs no greater than 0.5–1.0 cm in diameter. A routine maintenance programme avoids having to cut large limbs and provides for growth renewal each year and reasonable cropping on all sides. In contrast, trees allowed to grow together between rows over a number of years will become unproductive and require more severe pruning into older wood. Generally, the more severe the cuts, the more vigorous and vegetative the regrowth will be, often requiring 2 years or more to return to normal crop levels on that surface. By then, the surface will require severe cuts again.

Topping is also done with mechanical saws mounted on adjustable arms. Topping improves light penetration into the upper canopy and decreases tree height to reduce the cost of harvesting and spraying operations. It also improves spray coverage. As stated previously, tree height should be no greater than twice the width between the outer perimeter of the tree canopies between rows for optimum interception of light (Wheaton *et al.*, 1978). Limiting tree height becomes increasingly important in high density plantings with very narrow between-row spacings. Generally, tree height is maintained at about 4–5 m.

Trees are topped in different patterns. Some growers prune straight across the top of the tree in a flattop. Others set the cutting angle of the saws between 10 and 25° creating a rooftop effect. These patterns also remove varying amounts of wood from the sides of the tree. It is preferable to reestablish required tree height by making small cuts over several seasons rather than cutting into large wood. With hedging, removing large limbs produces vigorous, vegetative shoots from few growing points, thus defeating the purpose of topping. This is especially true for inherently vigorous scion cultivars such as lemon or for scions on vigorous rootstocks such as rough lemon.

GROWTH REGULATORS

Growth regulators have been used on a limited scale worldwide since the 1940s (Davies, 1986b) primarily to improve size and quality of fruit for the fresh market but also to prevent sprouting of small trees (Coggins, 1981; Wilson, 1983). The growth regulators, 2,4-D and gibberellic acid (GA_3), are applied routinely to navel oranges in California (Coggins, 1981) to improve peel quality and prevent fruit drop. Although the same materials have proved effective in other countries, they are not widely used because their effectiveness may vary from season to season and margin for error is small. Growth regulators are generally applied at very low concentrations (mg l^{-1}) and if improperly used may cause leaf, fruit or tree damage. Therefore, few citrus growers use growth regulators as part of their cultural programme. Although some increase in use of growth regulators may occur, it is unlikely that their use will become universal.

Growth regulators are used on a very limited basis to improve fruit set of mandarin-hybrids such as 'Orlando' tangelo, and 'Nova' and 'Robinson' tangerines. These hybrids are weakly parthenocarpic and without sufficient cross-pollination do not usually produce a commercial crop. Gibberellic acid applied at bloom increases fruit set and yields and produces seedless fruit which is especially desirable for the European market.

Many mandarins and mandarin hybrids produce large crops of small fruit followed by small crops of large fruit. This problem, termed 'alternate bearing', may cause marketing and management difficulties for growers because fruit less than a minimum size are unmarketable. Dilute sprays of naphthalene acetic acid (NAA) or ethephon after initial fruit set at the beginning of the 'physiological drop' period remove some of the fruit allowing the remaining fruit to grow to a more favourable size. Ethephon and/or NAA have been used successfully to thin mandarin crops in Japan, Israel, Florida and Australia (Coggins, 1981; Davies, 1986b). This also helps even out the crop load from year to year. It is important to thin the crop only during 'on' years and to avoid spraying trees of low vigour or that are stressed (Wheaton, 1981). Light hedging of mature trees after the postbloom drop period is another method for adjusting crop load in the on year.

Citrus fruits held on the tree past their normal harvest time typically soften, become off-coloured, and are prone to drop, thus reducing yields. Extension of the harvest season allows the grower to market fruit over a longer period and to take advantage of high prices for late-season fruit. Application of GA_3 to delay peel softening and colour development and 2,4-D to delay fruit drop has been used with success to extend the harvest season of grapefruit in Texas, Florida and South Africa and navel oranges in California and Australia, etc. (Coggins, 1981; Davies, 1986b). These materials also delay peel senescence and onset of peel breakdown known as 'black eye' of 'Minneola' tangelos stored on the tree. Spray application of 2,4-D alone delays premature fruit drop of 'Pineapple' oranges and decreases the severity of summer fruit drop of navel oranges in Florida (Davies, 1986b). It is important to consult current spray guides concerning limitations and label restrictions before using 2,4-D. Some crops such as peppers and grapes are damaged by very low levels of 2,4-D. Similarly, GA_3 applied at high rates defoliates citrus trees and causes fruit burn. Therefore, caution should be exercised in applying these materials.

6

WEEDS, PESTS AND DISEASES

Weeds, arthropods, nematode pests and pathogens are major constraints to citrus production in all areas of the world. Some problems have resulted in almost complete economic loss (greening in several Asian countries) and in other cases these problems have required major alterations in the industry (Brazil, Spain and Venezuela in response to tree losses from citrus tristeza virus [CTV]). In every country, managing these problems requires accurate information, proper strategies and production practice inputs in order to maintain economic yields.

WEEDS

Competition between weeds and mature citrus trees for nutrients and water is less severe than for younger trees. Nevertheless, severe weed pressure may reduce yields (Jordan, 1981), impede harvesting operations, and clog drainage ditches requiring special control (Vandiver, 1992a,b). Vines are a particularly acute problem because they may cover the entire canopy, thereby reducing light interception and hampering harvesting operations (Wondima-gegnehu and Singh, 1989). Control of weed growth is a major source of expense especially in tropical areas and southern areas of Florida, where high temperatures and rainfall support excessive weed growth.

An economically successful weed control programme depends on proper identification of weed species and selection of the most efficient control practices (De Barreda, 1977; Ito et al., 1981). Variation in programmes occurs from one area to another depending on orchard structure, soil conditions and irrigation system.

163

Table 6.1. Weeds commonly found in citrus orchards in the United States. Many of these species are also found worldwide, but this is by no means a complete list.

Common/family name	Scientific name	Life cycle	Characteristics
Grasses (*Gramineae*)			
Paragrass	*Panicum purpuresceus*	Perennial	Reproduce by stolons
Alexandergrass	*Brachicria plantaginea*	Annual	Reproduce by seeds
Guineagrass	*Panicum maximum*	Perennial	Reproduce by stolons
Vasseygrass	*Paspalum urvillei*	Perennial	
Bahiagrass	*Paspalum notatum*	Perennial	Seed propagated
Bermudagrass	*Cynodon dactylon*	Perennial	Seed, stolon, rhizome propagated
Torpedograss	*Panicum repens*	Perennial	Reproduce by rhizomes
Foxtail	*Setaria* sp.	Summer annual	Seed propagated
Texas panicum	*Panicum texanum*	Summer annual	Seed propagated
Sandspur	*Cenchrus pauciflorus*	Summer annual	Seed propagated
Maidencane	*Panicum hemitoman*	Perennial	Semiaquatic, hollow stems
Johnson grass	*Sorghum halapense*	Perennial	Reproduce by rhizomes
Brown top panicum	*Panicum fasciculatum*	Summer annual	Reproduce by seeds
Sedges (*Cyperaceae*)			
Yellow nutsedge	*Cyperus esculentus*	Perennial	Reproduce by seeds, rhizomes, tubers, triangular stem
Purple nutsedge	*Cyperus rotundus*	Perennial	Reproduce by seeds, rhizomes, tubers, forms tuber chains
Compositae			
Spanish needles	*Bidens bipinnata*	Summer annual	Flowers with yellow centre, white petals
Horseweed	*Erigeron canadensis*	Annual	Small yellow flowers, tough stems
Ragweed	*Ambrosia artemisiifolia*	Summer annual	Hairy and slender stems

Table 6.1. continued

Common/family name	Scientific name	Life cycle	Characteristics
Dogfennel	*Eupatorium capillifolium*	Perennial	Purplish stems, thin leaves
Sowthistle	*Sonchus* sp.	Winter annual	Prickly leaves, yellow flowers
Gourds (*Cucurbitaceae*)			
Balsamapple	*Momordica charantia*	Perennial	Vine, fruit orangish berry
Morning Glory (*Convolvulaceae*)			
Field bindweed	*Convolvulus arvensis*	Perennial	Thick vine, pink seed pods
Morning glory	*Ipomoea* sp.	Perennial and annual	Vines, funnel-shaped flowers in red, purple, white
Nightshades (*Solanaceae*)			
Brazilian nightshade	*Solanum seaforthianum*	Perennial	Vine, red berries
Milkweed (*Asclepiadaceae*)			
Milkweed vine	*Morrenia odorata*	Perennial	Vine, milky sap, seed pods with silky hairs
Vervain (*Verbenaceae*)			
Lantana	*Lantana camara*	Perennial	Woody shrub, pink, cream, purple and orange flowers
Pokesweed (*Phytolaccaceae*)			
Pokesberry	*Phytolacca americana*	Perennial	Reddish-purple, large taproot, purple fruit
Pigweed (*Amaranthaceae*)			
Redroot pigweed	*Amaranthus retroflexus*	Summer annual	Small green flowers, black seeds
Cashew (*Anacardiaceae*)			
Brazilian pepper tree	*Schinus terebinthifolius*	Perennial	Shrub or small tree, bright red fruit

Classification of Weeds

Accurate identification of weed species is necessary before selecting and implementing a control programme (De Barreda, 1977; Tucker and Singh, 1992). Weeds are classified on the basis of their structure, life cycle and seasonality. Grasses (monocotyledons with strap-shaped leaves) that cause problems in citrus orchards are generally perennial and reproduce by seeds and rhizomes (underground stems), e.g. Bermuda and torpedo grass, or vegetatively by tubers, stolons (above-ground stems) or bulbs. Sedges are similar morphologically to grasses but differ in having triangular rather than round stems. Sedges may be annual or perennial and reproduce by seeds, rhizomes and tubers also. Yellow and purple nutsedge are particularly troublesome to control because they are perennial and have an extensive system of rhizomes which produce underground tubers. Control measures for grasses and sedges may differ especially in choice of herbicides. Therefore it is important to distinguish between these species prior to selecting a control practice (Wondimagegnehu and Singh, 1989).

Broadleaf weeds are dicotyledons characterized by a branching growth habit and broad, netted leaves. Flowers of broadleaf weeds are generally more brightly coloured and more distinctive than those of grasses and have easily distinguishable parts (sepals, petals, ovaries, stamens).

Broadleaf and monocot weeds may be annuals, biennials or perennials. Annuals complete their life cycle from seed within a season. Summer annuals begin growing in the spring and produce seed and die during the late fall or winter. In contrast, winter annuals germinate at cooler temperatures in the fall and grow throughout the winter, usually dying during the summer as temperatures increase. Biennial species develop from seed during the fall and produce roots and leaves during the first season. Plants flower, produce seed and die by the fall of the second year. Perennial weeds live more than 2 years, either initiating growth in the spring (torpedo grass, nutsedge), or growing during the fall and becoming dormant during hot summer months (Florida betony). Many perennials accumulate reserves in rhizomes, tubers or stolons during their respective growing seasons becoming dormant and over-wintering or oversummering as these structures. Other perennials are woody and overwinter on stored carbohydrates or are evergreen.

More than 100 species of weeds have been recorded in citrus orchards in the United States, although fewer than 20 are of economic importance in mature orchards, depending on location and history of the site (Table 6.1). In the Mediterranean region (De Barreda, 1977; Lo Giudice, 1981b) and Japan (Suzuki, 1981; Ito et al., 1981) many of the problem weeds are of the same genus but different species than those found in the United States.

Weed Control Practices

Mechanical weed control is commonly used worldwide. Discing and clean cultivation was commonly practised for many years in the Ridge area of Florida. Some growers felt that regular discing not only controlled weed growth, but also chopped tree roots driving them deeper into the soil, thus improving drought tolerance. However, this so called 'dust mulching' actually damages surface roots and causes soil compaction even in sandy soils. Moreover, regular movement of equipment can cause mechanical damage to limbs and remove fruit from lower parts of the tree. There is no evidence that regular discing improves drought tolerance or root growth and, in fact, it may lessen tree vigour. Orchards kept weed-free with chemical herbicides are warmer than those having a sod cover during radiation freezes because of more heat flux out of the warmer soil. Discing is not used as a method of weed control in bedded orchards because bed maintenance requires stable soil and roots are located close to the soil surface so may be subject to mechanical damage. Clean cultivation may be disadvantageous also because of loss of habitat for beneficial insects necessary for biological pest control.

Mechanical mowing is commonly used as an effective method of weed control worldwide, particularly where soil erosion is a problem. A number of different types of mowers are available that operate either by direct drive or more commonly by a power take-off (PTO) from the tractor. Specially modified mowers are used for bedded orchards to permit mowing of the ditches and sides of the beds. These are called 'bat wing' mowers because they are hinged in the middle allowing for flexibility of motion. Mowing of perennials removes tops and reduces stored reserves, thus weakening the plant and reducing subsequent growth. Annual weeds should be mowed before they go to seed.

Chemical control

Chemical weed control has become a common practice in many citrus-growing regions. While chemical control is very effective if used properly, improper selection of materials, variations in environment, and improper application may result in either poor weed control or damage to citrus trees. Most important to any herbicide programme is to understand the chemical properties and limitations of the product. Product labels should be read and the instructions followed and carried out by the operator during application. Materials should be applied within the recommended ranges on the label. The concept that if a little is good, a lot is better is always dangerous when applying herbicides or other pesticides. Herbicides are applied before weed emergence (preemergence), or after the weeds emerge (postemergence). Some materials are applied before the trees are planted, but most are used after planting. Soil fumigation controls weed seed germination and may be considered a preplant application. Use of preemergent soil residual herbicides is quite common in mature orchards. These products should be applied before weeds have emerged or just as they are emerging in cases where some postemergence capability exists. Preemergent materials should be applied to a clean soil surface and require rainfall or irrigation to become effective. Ultraviolet rays from the sun may inactivate soil residual herbicides if they are not incorporated to a shallow level. Deep discing following application of residual herbicides defeats the purpose of chemical weed control by breaking up the herbicide layer and brings weed seeds to the surface as the herbicide is turned under.

Postemergence herbicides are applied after weeds have emerged and act either by contact or root uptake. Many postemergence materials are contact herbicides and thus are only effective if they come in contact with foliage or stems; they require good coverage because they are not translocated, and for this reason they also have no activity in the soil. Some are systemic and translocated to other plant parts including the roots. The various kinds of herbicides approved for use on citrus in each country should be available from a local advisory group.

Herbicides may be broadcast over an entire area, banded in a specific area, or directed to a particular region. Broadcast, also called trunk-to-trunk, herbicide application, is a common practice in many orchards, with chemical weed control replacing discing as a means of maintaining a clean orchard floor. Trunk-to-trunk programmes virtually eliminate all weed growth permitting ease of harvesting and spray operations. Amount used and cost of herbicides, problems of soil erosion and reduction of alternate habitat for biological control agents are major disadvantages of this system. Its future may be limited by the potential for groundwater contamination. Banded herbicide application in the tree row is used in some areas such as Florida and Brazil. A sod cover is maintained between rows with mechanical or chemical mowing. This system reduces herbicide costs, soil erosion and provides an alternative plant canopy for beneficial insects needed for biological control but provides less freeze protection and requires regular mowing. Directed or spot sprays are used primarily to control outbreaks of difficult weeds in particular areas of the orchard. In areas such as China, intercrops like cabbage, wheat or asparagus are grown between tree rows to maximize land use.

Establishment and maintenance of a weed control programme is an essential part of orchard practices. Well-designed programmes rely on a combination of methods and materials, rather than a single strategy. Programmes vary with location and severity of weed pressures and are more difficult in high rainfall and temperature areas (tropics). Because of increased herbicide costs in these areas, more hand cultivation is practised and weed control is frequently inadequate. Hand cultivation does not usually kill the reproductive root or stem parts of persistent grasses which quickly become reestablished.

Hand cultivation is not practical on ditch banks and in reservoir ponds typical of high rainfall, bedded orchard areas as found in southern Florida, Belize and other lowland tropical areas. Several herbicides are available for control of aquatic weeds (Vandiver, 1992a,b). Proper application is essential when applying herbicides to waterways or reservoirs because of the potential for environmental damage.

Many factors including temperature, rainfall, wind, soil type and stage of development of weed species influence herbicide efficiency (Wondimagegnehu and Singh, 1989). Temperature affects weed growth rate and uptake of herbicides. Low soil temperatures decrease uptake of residual materials and low air temperatures slow uptake and translocation of foliar applied materials. Soil residual herbicides usually require water from rain or sprinkler irrigation for activation and uptake. Low soil moisture induces water stress in the weed species, thereby decreasing foliar uptake and translocation of foliar-applied herbicides. Rainfall within 2 to 4 h of application of contact herbicides will greatly reduce or eliminate the effectiveness of these materials. Application of herbicides during windy conditions is inadvisable because of the risk of tree injury from spray drift, but also because of uneven distribution to the target weeds.

Soil type has a pronounced influence on the effectiveness of soil residual herbicides. Organic matter may complex herbicides reducing their effectiveness. In contrast, herbicides are readily available in the soil solution of coarse sandy soils and this may result in phytotoxicity to the tree. In addition, residual materials are readily leached from these soils resulting in noneffective levels of the herbicide while the leached material can contribute to surface- and groundwater pollution. Effective weed control using soil residual materials is especially difficult in orchards with highly variable soil types as found in the flatwood bedded orchards in Florida.

Stage of development of the target weed species has a major influence on performance of the herbicide. Most preemergent materials must be taken up by weed seedlings as they begin to grow and these materials become ineffective at later stages of weed development. Some herbicides require clean soil conditions to be effective. Weed cover intercepts the material before it reaches the soil. Initiation of a chemical weed control programme should start with a contact herbicide or clean cultivation before applying a preemergence soil residual material. Most contact herbicides are more effective on young, rapidly expanding tissues than older tissues that have developed thick, protective cuticles or pubescence. Some difficult-to-control perennials like brambles should be treated with herbicide during the autumn when materials are actively translocated to the roots. Application of herbicide is much less effective if made during the flowering period of the weed. Biological or chemical weed control materials should be applied to vines before emergence when the stems are not yet woody. Once vines have grown into the tree canopy, chemical control becomes almost impossible and they must be physically removed, a costly and difficult means of control. For these reasons, shallow cultivation or mowing before application of herbicides to heavy weed populations is often helpful.

Herbicide injury

Nearly all herbicides are potentially hazardous to citrus trees if not properly applied. Symptoms of herbicide injury, however, are not always obvious and may be confused with nutrient disorders, diseases or mechanical damage. The type of injury depends on the amount and kind of material applied. Contact herbicides produce localized damage to a particular part of the tree, usually near ground level where spray comes into contact with tree leaves. Rarely will symptoms be observed in the upper canopy of the tree unless a major error has been made. Damage may appear as small yellow or necrotic specks on leaves or as dieback of the entire limb if heavier contact is made. Damage from contact sprays is normally randomly distributed in an orchard since the herbicide applicator will not usually apply the material to every tree. Large trees with thick bark are not usually killed by improper application of contact herbicides to the trunk. The outer bark is still living and heavy application, particularly of systemic herbicides, on the trunk even at 5 to 10 years of age can girdle the tree.

Slight overapplication of residual herbicides is not easy to detect as leaf symptoms are not specific. However, in more severe cases, distinctive leaf chlorosis and even abscission may occur. Trees appear stunted and new leaves may be strap-shaped or lanceolate. Damage from residual herbicides generally occurs in blocks of trees or in entire rows, often associated with changes in soil type, irrigation practices, or an application rate error. This is why it is extremely important to follow label instructions carefully and to adjust rates based on soil conditions. Recent studies suggest that some soil-applied fungicides that improve root growth may also increase uptake of herbicides causing severe tree damage.

Sprayer calibration

Although many innovations in herbicide sprayer design have been introduced, proper and repeated calibration of herbicide sprayers is still the most important operation to achieve effective weed control, limit damage to trees and the environment from over application and save money. In any herbicide spray programme the goal is to apply a specific amount of material per treated area. Ground speed, spray nozzle operating pressure and pattern of output regulate the amount of herbicide applied (Matthews, 1992).

Pattern of spray application also affects output. Spray nozzles are manufactured to apply material in a number of different patterns and are rated in l (or g) min^{-1} at a given pressure. Commonly used nozzles include flat fan, wide-angle fan and even-spray patterns. The particular characteristics of a nozzle are stamped on it and additional information on output at various pressures is available from the manufacturer. The nozzle angle is chosen to allow adequate overlap of the spray at a given boom height and nozzle spacing. The optimum height is determined by the minimum tree skirt height, weed height and pressure to direct the spray the distance required along the outside of the spray pattern. Booms usually allow for independent shut-down of one or two nozzles near the tractor so that narrower strips can be treated for younger trees. Nozzles will wear with time causing an increase in output. For this reason they should be recalibrated at regular intervals and replaced when they exceed specified output by more than 5%.

Calibration should be done regularly after determining type of nozzle, operating pressure and boom height. Calibration may be done by spraying a prespecified area of ground or by operating the sprayer in a stationary mode at a carefully controlled rate and pressure. Detailed information on sprayers, spraying and calibration should be obtained from in-depth presentations on these subjects.

An alternative method of application is to deliver the herbicides to the irrigated area under the tree as done in Israel (Oren and Israeli, 1977) and Florida (Wondimagegnehu and Singh, 1989). Herbigation is practised in conjunction with use of microsprinkler irrigation. This system works well in a dry climate like in Israel where weeds tend to grow and compete for fertilizer and water only in the irrigated area under the tree. In high rainfall areas such as Florida, supplemental herbicide treatments must be applied to the remaining ground area that has significant tree root exploration. This system may be better applied to young trees in their first 2 or 3 years of growth when the tree root zone approximates the wetting zone of a sprinkler. This system does not work well if weeds are allowed to develop any height since they will interrupt the sprinkler spray pattern.

Biological control

Biological methods may have potential for future integrated control of weeds. *Phytophthora palmivera* has been shown to control milkweed vine in Florida if applied at the early seedling stage (Tucker and Singh, 1992). Allopathy may also have a role. For example, several compounds in extracts from *Lantana camara*, a weed itself, suppress the growth of rye grass plants (Singh *et al.*, 1989). In Florida, this allopathic effect is problematic allowing *Lantana* to develop preferentially in the field compared with easier-to-control weeds. Insects that feed on specific weeds like *Lantana* may also have promise for future biological control (Habeck, 1977).

PESTS

Mites, insects and nematodes that attack citrus number in the hundreds (Talhouk, 1975). Many of them only occasionally cause economic damage. However, there are pests that feed on the roots, trunk, stems, leaves or fruit and injure the tree enough to cause economic loss. In Florida, more than 21 different mites or insects are reported to have caused economic injury for fresh fruit production (Albrigo, 1978). Many of the insects that are reported to require control because of economic damage are listed in Table 6.2, and are of concern in most major citrus production areas of the world. One reason for this wide distribution is man. As people have moved citrus from the areas of origin or older production areas to other parts of the world (see Chapter 1), pests and diseases have often been moved also. Because of large numbers of pests, chemical control is widely practised by citrus producers. Many individual orchard pest outbreaks are related to chemical upsets of natural enemies. Biological control, which is cost-effective when it works and helps meet environmental and health concerns regarding use of pesticides on food products, is a major emphasis of current entomological research. Biological control has been practised in citrus for many years (Browning, 1992). Much of the current biological control has been developed from indigenous or unplanned introduction of control agents and researchers have learned to manage these systems efficiently. Many of the natural enemies were probably imported with citrus just as the pests were. Purposeful importations of predators and parasites have occurred also (McCoy, 1985). Most of the useful biological control agents are found in the areas of origin of the plant host (Chapter 1). The reader should refer to the World Crop Pest Series (Elsevier) or similar current reviews for detailed information concerning many of the mites and insects affecting citrus and plants in general.

Mites

Mites can be pests of citrus throughout the world. All mite damage results from feeding by piercing and sucking mouthparts. The smallest mites, *Eriophyidae* spp., cause some of the most pronounced injury. The citrus bud mite (*Aceria sheldoni* Ewing) causes malformed twigs, leaves and fruit from feeding on flowers and buds. Populations build to injurious levels more often on lemons than other citrus and malformed fruit often reach maturity. Under southern California conditions, the life cycle requires from about 10 d in the summer to 1 month in the winter. Mites must be observed with a hand lens during population increases in order to verify the mite as a cause of injury. Citrus bud mite causes problems on fresh citrus in the Mediterranean region, southern Africa and California.

The citrus rust mite (*Phyllocoptruta oleivora* Ashm.) occurs worldwide and

Table 6.2. Major citrus pests identified in various citrus production areas worldwide.

Order/family	Species	Common name	Distribution
Acarina			
Eriophyidae	*Aceria sheldoni* Ewing	Citrus bud mite	A, CAL, MED, SAF, SAM
	Phyllocoptruta oleivora Ashm.	Citrus rust mite	A, MED, NAM, SAF, SAM
Tetranychidae	*Eutetranychus anneckei* Meyer	Texas citrus mite	FL, IND, ISR, SAF, TEX
	Panonychus citri Mc G.	Citrus red mite	A, CAL, FL, MED, SAF, SAM
	Tetranychus spp.	Carmine spider mite	A, CAL, MED, P, SAF
Phytoptipalpidae	*Brevipalpus californicus* Bks.	Citrus flat mite (Z), false spider mite (US)	A, CAM, MED, NAM, SAF
	(*Brevipalpus australis* Tucker)		
	Brevipalpus phoenicis Geijskes	Broad mite, reddish black flat mite	A, B, CAM, MED, NAM, SAF
Thysanoptera			
Thripidae	*Frankliniella occidentalis* Perg.	Western flower thrips, Grass thrips	ARIZ, CAL, SAM, SPAIN
	Heliothrips haemorrhoidalis Bché.	Greenhouse thrips	A, CAM, FL, MED, SAF, SAM
	Scirtothrips auranti Faure	Orange (Citrus) thrips	SAF
	Scirtothrips citri Moult.	Citrus thrips	ARIZ, CAL, CAM, P
	Thrips tabaci Lind.	Onion thrips	A, CAL, CAM, MED, SAF, SAM
Hemiptera-Homoptera			
Aleyrodidae	*Aleurocanthus spiniferus* Quaint.	Orange (Citrus) spiny whitefly	A, CAM
	Aleurocanthus woglumi Ashby	Citrus blackfly	A, CAM, FL, SAF, TEX
	Aleurothrixus floccosus Mask.	Flocculent (Wooly) whitefly	A, CAM, MED, SAF, SAM
	Bemisia citricola Gomez-Clemente	Whitefly	A, MED
	Dialeurodes citri Ashm.	Citrus whitefly	A, CAM, CHILE, FL, MED
	Dialeurodes citrifolii Morg.	Cloudy winged whitefly	B, FL, IND

Table 6.2. continued

Order/family	Species	Common name	Distribution
Aphididae	Aphis gossypii Glov.	Cotton aphid	A, CAM, MED, NAM, SAF, SAM
	Aphis spiraecola Patch	Spirea aphid	CAM, MED, NAM, SAM
	Myzus persicae Sulz.	Green peach aphid	A, CAM, MED, NAM, SAF, SAM
	Toxoptera aurantii Boy.	Black citrus aphid	A, CAM, MED, NAM, SAF, SAM
	Toxoptera citricida Kirk.	Brown citrus aphid	A, CAM, SAF, SAM
Margarodidae	Icerya purchasi Mask.	Australian fluted scale, Cottony cushion scale (US)	A, CAM, MED, NAM, SAF, SAM
Pseudococcidae	Planococcus citri Risso (Pseudococcus commonus)	Citrus mealybug (US), Common mealybug	A, CAM, MED, NAM, SAF, SAM
	Pseudococcus adonidum L.	Long-tailed mealybug	A, B, MED, MEX, SAF
	Pseudococcus citriculus Green	Citrus mealybug (Japan)	A, EGY, ISR
	Pseudococcus comstocki Kuw.	Comstock mealybug	A, B, CAM, EGY, ISR
	Ceroplastes floridensis Comst.	Florida wax scale	A, AUS, B, FL, MED, MEX
Coccidae	Ceroplastes destructor (Newst)	White wax scale	AUS, SAF
	Ceroplastes rubens Mask.	Pink (Red) wax scale	A, AUS, ITALY
	Ceroplastes sinensis Del Guer.	Chinese wax scale	AUS, B, MED
	Coccus hesperidum L.	Brown soft scale	A, AUS, CAM, MED, NAM, SAF, SAM
	Coccus viridis Green	Green coffee scale	A, CAM, SAM
	Saissetia hemisphaerica Targ. (S. coffeae Walk.)	Hard brown scale, Coffee helmet scale	A, CAM, MED, SAM
	Saissetia oleae Bern.	Black scale (US), Brown olive scale (A)	AUS, CAM, IND, MED, NAM, SAF, SAM
Diaspididae	Aonidiella aurantii Mask.	California red scale	A, SAF, MED, SAM, CAM, NAM, AUS

Taxon / Species	Common name	Distribution
Aonidiella citrina Coq.	Yellow scale	AUS, CAL, FL, IND, JAPAN, MEX
Chrysomphalus aonidum L. (*Chrysomphalus ficus* Ashm.)	Florida red scale (US), Circular purple scale	A, AUS, CAM, MED, NAM, SAF, SAM
Chrysomphalus dictyospermi Morg.	Dictyospermum scale	MED, SAM
Lepidosaphes beckii Newm.	Purple scale	CAM, MED, NAM, SAF, SAM
Lepidosaphes gloverii Pack. (*Insulaspis g.*)	Glover scale (US), Long mussel scale (Z)	B, CAM, JAPAN, MED, NAM, SAF
Parlatoria pergandii Comst.	Chaff scale	A, CAM, MED, NAM, SAF, SAM
Pinnaspis aspidistrae Sign.	Aspidistra (Fern) scale	A, FL, SAM
Selenaspidus articulatus Morg.	Rufous scale	CAM, SAM, SEA
Unaspis citri Comst.	Snow scale	AUS, CAM, CHINA, FL, SAM

Heteroptera

Pentatomidae

Rhynchocoris humeralis Thunb.	Citrus green stink bug	A, MED

Coreidae

Leptoglossus phyllopus L.	Leaf-footed bug	FL, MEX, SAM

Coleoptera

Cerambycidae

Chelidonium gibbicolle White	Roundheaded borer	CHINA
Melanauster chinensis Först.	Black and white citrus borer	CHINA, JAPAN

Curculionidae

Otiorrhynchus cribricollis Gyll.		
Diaprepes spp.	Sugarcane root stalk borer	CAM, FL, SAM
Pachnaeus citri Mshl.	Citrus root weevil	CAM, FL
Pantomorus cervinus Boh.	Fuller rose beetle	NAM, SAF, SAM

Hymenoptera

Formicidae

Acromyrmex octospinosus Reich.	Leaf-cutting ant	A, CAM, SAM
Atta cephalotes L.	Leaf-cutting ant	B, CAM, PERU
Atta sexdens L.	Leaf-cutting ant	CAM, SAM
Solenopsis invicta Buren	Imported fire ant	CAM, FL

Table 6.2. continued

Order/family	Species	Common name	Distribution
Lepidoptera			
Lyonetiidae	*Phyllocnistis citrella* Staint.	Citrus leaf miner	A, AUS
Hyponomeutidae	*Prays citri* Mill.	Citrus flower moth, bud moth	A, MED
Metarbelidae	*Indarbela tetraonis* Moore	Litchi stem borer	A
Papilionidae	*Papilio demoleus demoleus* L.	Citrus leaf-eating caterpillar	A
	Papilio memnon L.	Pastor swallowtail	A
	Papilio polytes polytes L.	Citrus butterfly	A
Diptera			
Trypetidae	*Anastrepha fraterculus* Wied.	South American fruit fly	CAM, SAM
	Ceratitis capitata Wied.	Mediterranean fruit fly	AUS, CAM, MED, SAF, SAM
	Dacus dorsalis Hend.	Oriental fruit fly	A
	Dacus tsuneonis Miyake	Chinese citrus (Japanese orange) fruit fly	CHINA, JAPAN

Abbreviations: A, Asia; ARIZ, Arizona; AUS, Australia; B, Brazil; CAL, California; CAM, Central America and Caribbean; EGY, Egypt; FL, Florida; IND, India; ISR, Israel; MEX, Mexico; NAM, North America; P, Pakistan; SAF, southern Africa; SAM, South America; SEA, southeast Asia; TEX, Texas.

can be a pest of fresh fruit production in most humid climate production areas. Populations of this mite usually increase faster on fruit and can develop densities of 70–100 mites cm^{-2} on the fruit surface before injury occurs (McCoy and Albrigo, 1975). However, some hybrid cultivars like 'Sunburst' mandarin are a preferred host and citrus rust mites can increase to higher populations on leaves than fruit. This cultivar may be more sensitive to injury as leaf injury can develop into severe blister-like lesions on the leaves and young stems (Albrigo et al., 1987). Because mites have small mouthparts, feeding injury only extends to the epidermal cells and requires several feeding probes per cell over a short period in order to kill cells. Subsequent damage development follows formation of a wound periderm under the injured cells and break-up and some sloughing of the injured layer. In late injury, no wound periderm forms and the injured cells remain intact but dark brown and the surface remains shiny (Albrigo and McCoy, 1974). Severe early damage, greater than 50–75% surface injury, can lead to reduced fruit growth, excess drop and fermentation off-flavour development in late harvested fruit (Allen, 1978; McCoy, 1988)

In warm, humid subtropical areas where rust mite is the most severe problem, reproductive cycles require 7 to 14 d depending on temperature. Mite numbers increase in late spring or early summer and damage often occurs before the rainy period when the entomophagous fungus, *Hirsutella thompsonii*, develops, invades the mites and suppresses the population. Sufficient populations may again increase in the fall to injure fully developed fruit. Chemical control is usually required for fresh fruit production. In a dry Mediterranean climate, citrus rust mite control was adequate after the introduction and mass rearing of predaceous mites and the judicious use of selective miticides instead of sulfur, which greatly reduces populations of natural predators (Cohen, 1975). Under similar climatic conditions in California, selective chemicals integrated with other practices can successfully control citrus rust mite (California IPM Manual Group, 1984). Some other Eriophyoids such as the pink citrus rust mite (*Aculops pelekassi* [K.]) and the citrus blotch mite (*Calacarus citrifoli* K.) occasionally also cause fruit damage.

Other mites that cause injury similar to that due to the Eriophyoids are the broad mites, *Hemitarsonemus latus* Banks, *Brevipalpus californicus* Banks or *Brevipalpus phoenicis* Geijskes. *Brevipalpus phoenicis* is common and able to develop larger populations in the warmer climate of central Brazil (25.7°C max. and 12.8°C min. during the coldest 4 months) than in central Florida (23.7 max. and 11.7°C min.). *Brevipalpus* occasionally reaches injurious populations in south Florida. These small mites reproduce in 3–7 d. *Brevipalpus phoenicis* is also the carrier of the pathogen for the disease leprosis, which occurs in Brazil but not in Florida (see p. 198).

The spider mites (Family Tetranychidae) are broadly distributed, with the two common pest species being the citrus red mite (*Panonuchus citri* McGregor) and the Texas citrus mite (*Eotetranychus banksi* [McG.]). These species and the carmine spider mite (*Tetranychus cinnabarinus* Boise) rapidly develop injurious populations in early spring, damaging leaves and in some cases causing defoliation. These mites have not been observed to cause russet injury to fruit (Albrigo *et al.*, 1981). The six-spotted mite (*Eotetranychus sexmaculatus* [Riley]) only injures leaves from small colonies on the underside of the leaf. Most spider mites feed to a depth of about 100 µm and completely remove cell contents so that no oxidized cytoplasm can develop. These areas appear translucent due to the cells void of cytoplasm. New cells are formed from a more central periderm and these cells displace the empty cells. When damage is severe, injury is not repaired and in conjunction with water stress, mesophyll collapse can occur. Development of this disorder first appears as chlorotic zones in the leaf blade, followed by leaf browning and finally premature leaf drop. Citrus red mite densities under 20 leaf^{-1} did not appear to have any adverse effect on tree productivity in California (Hare and Phillips, 1992) suggesting that leaf cell replacement rates are adequate to compensate for cell losses from feeding by populations of that size.

Spider mites, like citrus rust mite, reach higher populations on lemon, but all citrus are fed upon. One generation of the spider mite species requires about 2 weeks at optimum spring temperatures, 30°C max. and 19°C min., with 10 to 12 generations year^{-1}. They overwinter as adults and populations usually decline during summer rainy seasons in marginally subtropical climates like Florida due to biotic and abiotic factors. Although predaceous mites, such as *Euseius hibisci*, and lady beetles, *Stethorus picipes*, attack spider mites, the current predator complex in California and Florida does not usually control citrus red mite or Texas citrus mite, particularly when spray programmes of organic phosphates or sulfur upset natural control. Economic losses from spider mite feeding in field trees appears to be rare. Therefore, little additional work on biological control has occurred.

Insects

Thrips are widely distributed but appear to be primarily a problem for the fruit blemishes that they cause. They have modified mouthparts for rasping and sucking. The early flower activity of the western flower thrip (*Frankliniella occidentalis* Pergande) or citrus thrips (*Scirtothrips citri* Moult.), as occurs in California, often results in a ring of scarred peel tissue around the stem end of the fruit. Occasionally, population increases of greenhouse thrips (*Heliothrips haemorrhoidalis* Bche.), may occur on grapefruit in Florida, and injury may result in superficial scarring at the boundary where two fruit touch leaving a ring of injured peel, similar to injury that occurs from citrus rust mite feeding. Life cycles range from 27 to 10 d for temperatures of 16 to 30°C. Citrus thrips are known to scar fruit in late spring during the second generation in California (California IPM Manual Group, 1984). Orchards with ground cover and minimum spray programmes that avoid organophosphates tend to have fewer thrip problems. Predaceous mites, *E. hibisci* and *Anystis agilis*, and the minute pirate bug prey on thrips as do natural enemies found in ground litter. A single spray at petal fall will usually control thrips if other factors are favourable for a minimum population increase.

Whiteflies and blackfly are most serious in nursery tree production but buildup on young flushes is significant in that it leads to sooty mould development (see p. 192). Whiteflies have sucking mouthparts and feed on leaves. Cloudy winged (*Dialeurodes citrifolii* Morg.) and other whiteflies have become a problem in nursery trees, particularly greenhouse production in Florida and Brazil. This and other whiteflies are difficult to control in the greenhouse because frequent sprays are required. Citrus and woolly whiteflies are controlled under Mediterranean climates by parasitic wasps such as *Cales noacki* and *Eretmocerus* spp. This control is easily disturbed by ants protecting this source of honeydew. Dust or insecticide treatments such as sulfur also kill these wasps.

The citrus blackfly (*Aleurocanthus woglumi* Ashmead) was a serious pest in Mexico and moved into Texas in the early 1970s and Florida in 1976. Blackflies feed on the leaves and produce honeydew that contributes to a severe accumulation of sooty mould and a decrease in tree vigour. Successful biological control of blackfly occurs when the two parasitic wasps (*Encarsia opulenta* and *Amitis hesperidum*) are present. These parasites are easily killed by sprays of organophosphates or sulfur, while the blackfly is difficult to kill with insecticides. Best control is maintained by introduction of the parasites and avoidance of these chemicals (Knapp and Browning, 1989).

Aphids also have piercing-sucking mouthparts and are phloem feeders but are usually not economic pests *per se*. They are important vectors of CTV and possibly other viruses. The brown citrus aphid (*Toxoptera citricida* Kirk.) is the most efficient vector and is capable of developing populations that can cover a spring flush on young trees and probably reduce tree growth (Lee *et al.*, 1992). These populations increase very fast on both leaves and stems of the spring flush, with reproduction time as short as 4 d in the warm spring but up to 10 d in cooler periods. A number of parasites, predators and fungi have been observed associated with aphids in South and Central America, some of which may control the brown citrus aphid. Biological control agents usually result in decreasing the levels of a pest rather than eradication. For an efficient virus vector like the brown citrus aphid, this level of control would probably only slow down the spread of CTV (see next section).

More kinds of scale insects attack citrus than any other family of insects. Because scale insects are sessile, small and often inconspicuous, they have been spread widely on citrus plants and are now present in most citrus production areas of the world. Some scales are mobile throughout all stages of development, but many become immobile under soft or armour covers after the crawler stage. Eggs are protected under the cover or the body of the female. A few to a few thousand eggs are produced by each female. Winged males are found in most species, but reproduction is parthenogenic in others. Two to ten generations are produced each year depending on the species. From crawler to adult there are as many as three stages in females and the long stylet mouthparts of all stages can be found in the plant tissues below the resting adult. Most efficient scales penetrate to phloem cells for long-term feeding.

Fluted scales like the cottony cushion scale (*Icarya purchasi* Mask.) produce a copious waxy deposit over the body. This is the most common and widely spread of the fluted scales. Up to 1000 eggs may be carried in the waxy mass. Dense colonies can occur on branches and the trunk. Infested small branches and fruit stems are often weak and fruit abscission can occur.

Mealybugs have an appearance similar to the fluted scales but with less wax. The citrus mealybug (*Planoccocus citri* Risso) is the most widespread species. This insect inhabits the stylar cavity of navel oranges and the area around the calyx which can lead to yellowing of the fruit and subsequent abscission.

The soft scales include several citrus pests like *Ceroplastes* spp. (white wax scale) and *Saissetia* spp. (black scale and brown soft scale). Generations per year range from two or three for the lowest fecundity to four to ten generations per year for brown soft scale (*Coccus hesperidum* L.). In all cases, more generations are produced in tropical climates and in warmer years. Maritime locations are also more favourable because of moderated climates. Young crawlers can be found on all parts of the canopy, but adults tend to settle on twigs and to a limited extent in the protected area near the midrib of the leaves. Preferred plant part varies with species. These insects also produce honeydew which attracts ants and causes sooty mould development.

Scale insect pests of citrus are generally under some of the best biological control of any insects (McCoy, 1985; Browning as reported by G. Smith, 1992). Soft scales like brown soft scale and black scale are parasitized by *Metaphycus* spp., but the host parasites are usually specific and pest species often differ by location. Ladybird beetle larvae and parasitic flies also contribute to soft scale control. The most common cause of upset of these control equilibria is the use of broad spectrum pesticides that kill predators and parasites as well as the target pest. Programmes with spray oil and specific target pesticides are essential for effective IPM programmes.

Armoured scales are protected by their wax-based covering. California red scale (*Aonidiella aurantii* Mask.) is an important and damaging member of this family. The first generation crawlers rapidly disperse even to small fruit causing noticeable distortion. Infested young leaves become yellow and small twigs may die. Two more generations are produced during the early and late summer, but infestation of more mature fruit does not lead to their being misshapen. Many successful parasitic wasps attacking scale insects are of the genus *Aphytis*. In California, *Aphytis melinus* is an important parasite of the

California red scale (California IPM Manual Group, 1984). Combined with the effects of *Comperiella bifasciata* (Howard), another effective parasite, moderate biological control is obtained. The Florida red scale (*Chrysomphalus aonidum* L.) and the dictyospermi scale have similar biologies to the California red scale, but they are much less severe pests. Florida red scale is controlled by *Aphytis holoxanthus* DeBach, which was purposely introduced (Bullock and Brooks, 1975). In South Africa, the predators *Cilocorus negrita* and *Cilocorus distigma* combined with *Aphytis* spp. give good control of California red scale.

Glover, chaff and citrus snow scale are less well-controlled biologically, but the first two are seldom a problem if repeated sulfur sprays are avoided and some ground cover is maintained to provide a good habitat for predators. Chaff scale (*Parlatoria pergandii* Comst.) is characterized by reservoir populations on limbs and trunks, but it can be widely distributed on the tree and feeds on leaves, stems and fruit tissue which sometimes leads to fruit abscission. This is a problem in some citrus-growing locations such as Brazil where clean cultivation is practised which eliminates the desirable higher humidity habitat for alternate hosts for the predators. Heavy use of inexpensive, broad-spectrum pesticides like sulfur also kill predators and parasites exacerbating the problem. Purple (*Lepidosaphes beckii* Newm.) and glover (*Lepidosaphes gloverii* Pack.) scales are very similar having two main reproductive cycles per year. These scales feed on all plant parts. Excessive feeding on leaves results in chlorosis. In Florida, purple scale is controlled by *Aphytis lepidosaphes*, a chance introduction.

Citrus snow scale (*Unaspis citri* Comst.) is also widely distributed but is primarily a limb and trunk feeder. Large populations can build up and turn the trunk and primary scaffolds white (males). The related species, *Unaspis yanonensis* Kuwana, referred to as arrowhead scale, is more severe on fruit and leaves and results in delayed colour development. This scale is a prominent problem on mandarins in Japan. *Aphytis lingnanensis* Compere was introduced to Florida for snow scale control, but it has been largely unsuccessful. Some predators (ladybird beetle from China being tested in Australia) and parasites appear to be better biological control agents for this difficult to spray scale insect.

Plant bugs (there are several species in Heteroptera) will feed on mature fruit, particularly if their natural weed host population is disturbed. This commonly occurs when discing or chopping is employed to prepare the orchard floor for harvesting (Albrigo and Bullock, 1977). Year-round availability of alternate hosts such as nightshade, guava, citrons, etc. should be avoided. Only the adults feed on fruit, and they penetrate into the juice vesicles with their long stylets. Generally there is little external damage, and most losses are due to decay development in the feeding punctures between oil glands. Feeding punctures through oil glands appear to be sterilized by the toxic oils. Sometimes small oil-induced necrotic spots are visible. Dry juice vesicles at the outer surface of the segments and a reddish reaction of these vesicles during alkaline peeling for sectionizing are sometimes the only indication of feeding.

Wood borers have not been a problem for citrus production in Europe or the American continent, but borers infest citrus in Japan, China (Talhouk, 1975) and Jamaica. Borers cause damage during the larvae stage using chewing mouthparts and usually bore into the trunk and primary scaffolds. They usually attack trees that are already weakened from other problems. Invasion of wood by secondary decay fungi often occurs. In the tropical Americas, however, root weevil larvae cause extensive damage to citrus, including death of young trees because of girdling of the trunk or primary roots from root bark feeding. These insects are also reported to cause injury of economic significance in Brazil and Florida (Futch and McCoy, 1992). Adults feed on and notch leaf edges, but overall have little direct impact on the tree except in young trees fed on by unusually large numbers of adults. The adults lay egg masses between two leaves cemented together or in other protected areas such as the fruit calyx area in the case of the Fuller rose beetle (*Asynonychus godmani* Crotch). Small larvae hatch in about 7–20 d and fall to the soil where they initially feed on fibrous roots. As they enlarge, larvae often move to larger roots and can girdle these. The larva stage may feed for only 35–40 d (*Artipus*) to 6–24 months (most other species). Damage from these insects is particularly severe in Florida and the Caribbean where the Fiddler beetles (*Prepodes* spp.) and sugarcane root stalk borer (*Diaprepes* spp.) predominate.

The weevil complex is one of the most difficult to control. A number of biological control agents and cultural practice strategies exist that together can give some level of control. Spray programmes have been fairly ineffective as discussed earlier. In the 1940s, it was found that citrus trees could be planted on mounds and then the primary roots exposed so that weevil larvae could not inhabit the area at the base of the trunk to girdle it and the small major roots. This strategy has been combined with the use of a tight mesh black polyethylene screen over the mound. When the larva fall to the ground from the hatched eggs, they can not penetrate the soil directly under the tree but instead take up residence on lateral feeder root areas at the edge of the canopy. This allows the young tree to become established since water can still penetrate the mesh. Adults are often collected by hand as they feed on the leaves. Entomophagous fungi, *Beauveria bassiana*, *Metarrhizium anisopliae* and other species, and *Steinernema* spp., attack the larval stage in the ground. Predators and parasites also exist but they have not provided consistent control.

Leaf cutting ants (*Acromyrmex octospinosus* Reich and *Atta* spp.) can cause severe leaf loss to the point that growth of young trees is inhibited, and trees become weak and are susceptible to foot rot and other fungi. Leaf pieces are removed and used as substrata on which to culture fungi in the ant colonies for food. Ant beds can cover large areas of the orchard. These ants are a serious problem in Brazil and South America in general. Imported fire ants (*Solenopsis invicta* Buren) have become a serious problem in new plantings in Florida. The ants colonize the soil near the trunk and feed on the tender trunk bark, particularly under freeze protection wraps. In some cases girdling occurs, but more often the trees are lost to infection by *Phytophthora* spp. Similar losses occur in recently cleared land in tropical America where the colonies develop in the feeder root area. Argentine ant species attack blossoms and transport and protect aphids, mealybugs, scales, whitefly, etc. from natural enemies. This increases honeydew production and associated sooty mould. Most baits and sprays have short residual effects or reinfestation occurs requiring continual monitoring and retreatment where ants are a problem.

Termites (*Reticulitermes flavipes* Kollor) also can cause tree losses under circumstances similar to those which promote fire ants. Young trees on recently cleared land are good hosts for these disrupted termite colonies (Stansly *et al.*, 1991). Direct tree death from girdling occurs, but some losses are due to phytophthora invasion through the feeding injuries.

Grasshopper populations can become large enough to cause defoliation of nursery stock or young trees in the orchard. Eastern lubbers (*Romalea microptera* Beauv.) and the American grasshopper (*Schistocera americana* Drury) are pests of economic significance some years in Florida.

Lepidoptera such as leaf miners, leafrollers and butterflies are generally not economic pests in most parts of the world. In some locations in Asia, species in this order are pests. The caterpillars are similar and are often referred to generically as orange worms. Occasional leaf or fruit damage occurs by leaf miners, leafrollers and butterfly caterpillars in most growing regions. In southeast Asia, Australia and very recently in Florida, citrus leaf miners are a serious pest. More damage is usually observed on young than mature trees. The orangedog butterfly (*Papilio cresphontes* Cram.) can be an economic problem in nursery production where so much new foliage is present and one or two caterpillars can defoliate a seedling or newly budded tree. Proportionally more tender flush, ideal for feeding, is usually available on very young compared with mature trees in the orchard and economic damage has usually been associated with young trees. Various Lepidoptera attack citrus usually on an inconsistent basis only making it difficult to develop any standardized control practices (California IPM Manual Group, 1984). Several predators and parasites exist and can be promoted by sound pesticide management so that these minor pests are even less of a problem.

A number of fruit flies (Diptera: Tephritidae) attack citrus throughout the world. Adults oviposit in mature fruit and the larvae feed and develop in the fruit pulp. Except in isolated instances, fruit losses are usually not high but fresh marketing is difficult because infested fruit cannot be easily graded out. Quarantines against importation of fruit from areas with various species of fruit flies significantly restricts movement of fresh fruit in world markets. The

Mediterranean fruit fly (*Ceratitis capitata* Weid.) is widely distributed in Europe, Africa and South America. It has a wide host range and will deposit eggs in relatively immature citrus. The South American fruit fly (*Anastrepha fraterculus* Weid.) and the oriental fruit fly (*Dacus dorsalis* Hend.) are widely spread within each respective continent. The Caribbean fruit fly (*Anastrepha suspensa* Loew) has been a problem for Florida grapefruit export shipments but rarely deposits eggs in grapefruit until the fruit is overmature. Other *Anastrepha* spp. exist with potential to harm citrus, especially in Central and South America. Current control stategies or shipping from fly-free zones are not very satisfactory from a long term point of view.

Nematodes

Nematodes are microscopic, cylindrical worms many of which are parasitic to plants and feed on fibrous roots. Three nematodes account for most of the problems experienced in citrus production around the world, but six genera are reported to attack citrus (Duncan and Cohn, 1990). The citrus nematode (*Tylenchulus semipenetrans* Cobb) is widespread and the most common nematode on citrus. It is a common problem in Florida and South and Central America, but also occurs throughout the Mediterranean Basin (Lo Guidice, 1981a). At least four biotypes exist worldwide. The nematode partially penetrates the feeder roots and establishes a permanent feeding site. Eggs are laid and remain near the posterior end of the adult. The larvae hatch in 12–14 d and migrate to new feeder roots nearby so that natural spread from tree to tree is fairly slow. Adults survive until normal root death which is probably not more than 1–2 years. Trees are generally weak without specific symptoms even though they may be on a good nutrition and irrigation programme. Under Florida conditions, this nematode causes more severe symptoms in bedded orchards. This may be due to the shallow root systems that are very sensitive to any stress.

The burrowing nematode (*Radopholus citrophilus* Huettel, Dickson & Kaplan) and *Pratylenchus* spp. are endoparasitic nematodes. They enter the root and can complete their life cycle there. Reproduction and growth are optimum at 24°C, but the nematode is active from 12 to 32°C. Larvae move to new roots in search of new food sources. Generally, the damage by an individual nematode is more severe than from a citrus nematode. The burrowing nematode is primarily a problem in Florida where it leads to spreading decline primarily on deep sandy soils. This may be due to the more severe damage to deep rather than shallow roots. Preferential feeding may occur due to more favourable soil temperatures below 30 cm or to less competition between nematodes and other soil-borne organisms at this depth. Nematode feeding leads to decline and dieback. All trees in the nematode zone decline and trees on the border of the affected area develop symptoms as the nematodes spread. *Pratylenchus* spp. are more widespread and some other nematodes may be of isolated importance in citrus areas. Most site-to-site spread is by man on new plant material (often ornamentals) or with cultivation equipment. Best control is by avoiding contamination of planting sites and complete preplant sterilization if required. Some rootstocks are tolerant of some nematodes, e.g. 'Milam' to burrowing nematode (Chapter 4).

Pest Management

Both chemical and biological methods are important in the control of the many pests of citrus and form part of an integrated pest management (IPM) programme. Chemical and biological control methods have several things in common. Successful methods of both types require efficacy. Properly timed control according to the biology of the pest insect and its interaction with the plant is important also.

Chemical control

Chemical control has evolved from the use of basic inorganic materials like the Bordeaux mix of lime-sulfur and copper sulfate applied by hand sprayers to currently using sophisticated application machinery and organic pesticides, some with very specific pest ranges and some that control specific development stages of the insect, i.e. those chemicals that specifically alter development by growth regulation (Knapp, 1992). Moults from one developmental stage to another are inhibited so that the life cycle can not be completed (Henrick, 1982). Specific chemicals are recommended in most major production areas, and information on these is published frequently since new chemicals, new rates and cancellations of chemicals occur frequently.

Application technology is as important as the choice of chemical. Timing and effective coverage are the two major considerations for proper application. Timing depends on the stage of insect development controlled by the pesticide, insect population dynamics, and the residual control period after application. If many generations of an insect are produced each growing season, then all stages of development will be present. Long-term residual control from the chemical is desirable to break the cycles of overlapping generations. Timing of the first application should be just in advance of the populations reaching an economic injury threshold, but these thresholds have not been established for most pests. Populations of economic pests usually increase very rapidly during peak development. Therefore, sufficient time must be allowed for spraying all orchards before injury occurs. Proper timing of control should be based on a scouting procedure to determine when a population increase is approaching injurious levels. An attempt to follow this procedure in Florida has been based on the observation that injury from citrus rust mite feeding occurs at population densities of 70–100 mites cm^{-2} (McCoy and Albrigo, 1975). Evolution of a mite–days to injury concept is presented in a Citrus Pest Management Guide (Knapp, 1992).

Pests with only two or three generations $year^{-1}$ need more precise timing in relation to development stage. Chemicals usually work on either the adult or larval stage and must be timed for when that stage predominates. This may be further complicated by where on the plant that stage of development can be found. Weevils provide an example of this complexity. The adults are leaf feeders, but the hatched larvae fall to the soil and burrow to roots where they feed. Neither stage is easy to spray directly and current chemicals have shorter residual times to protect the environment. This means that control lasts only a short time and many of the pests escape control from any one treatment.

The other important component of pest control is proper application. Aerial or ground spraying, granular ground spread and soil incorporation are used. Fumigants for nematodes and aldicarb are current examples of soil-incorporated materials. Because of degradation and toxicity these materials are usually drilled into the ground. Granular baits for ants are usually broadcast around the base of young trees or placed under tree wraps where freeze protection is required. Other special applications include application of parasitic nematodes for weevil larvae control using the microirrigation system or a herbicide spray boom followed by irrigation (Knapp, 1992).

Most other applications use ground or aerial spray equipment. Fixed wing and helicopter application are used in many countries. These methods are limited in their ability to deliver pesticides to the interior of the tree or even the underside of leaves. They are effective on pests that inhabit the outside of the canopy and upper surface of leaves like citrus red mite, but provide poor control of scale insects thay may cause heavy infestation on the trunk and limbs and greasy spot fungus that invades through stomata (see p. 194). An important advantage of aerial applications is that they can be made quickly to large areas of trees.

In the case of insects like snow scale that primarily inhabit the trunk and main scaffolds, hand spraying of infested trees is the most effective application method (Knapp and Browning, 1989). For good canopy penetration and moderate inner wood coverage, at least some form of ground, air carrier sprayer is needed.

The current procedure is to minimize the carrier solution (usually water) to cover more area per tank of spray and to minimize run-off that wastes material and increases potential soil contamination problems. This has been accomplished with traditional sprayers by using lower delivery volume nozzles and/or increasing ground speed. Newer sprayers are designed to create smaller droplets and the air carrier system is modified to carry these small droplets into the canopy to achieve good coverage with small volumes of spray (air curtain). Penetration is not as good with these sprayers. Most modern sprayers have electric eye sensors to evaluate tree height and spacing.

Nozzles are only turned on in response to sensing canopy at that height. Small trees or replants only receive a short burst as the sprayer passes. No spray is delivered to open spaces, which conserves pesticides and reduces environmental problems from spraying excess chemical on to the ground. Adequate dilute spray volume to cover the canopy depends on tree height and assumes full tree development in the tree row. Specific calibration is based on trees ha^{-1} (canopy volume) and therefore number of litres tree^{-1} that should be delivered.

Whichever machine or concentration is used, frequent calibration is important as with herbicide sprayers, for nozzles will wear or become clogged. It is also important to remember that most spray regulations now require that a limit of chemical per area is not exceeded (Knapp, 1992). Calibration and calculations for concentrate spraying must be done carefully so that product per area is not exceeded.

Spray equipment is expensive to buy and operate. Many of the more effective organic chemicals are also expensive. Methods that do not require or minimize spraying are desirable for cost reasons and because of environmental considerations. Integrated pest management is a realistic approach that properly maximizes use of biological control but recognizes the proper use of target specific pesticides when needed.

Biological control

Biological control has been practised for many years in citrus (Browning, 1992). In many cases, researchers discovered that biological control was already occurring and learned how to manage it more efficiently. Many of these natural enemies were probably imported with citrus plants just as the pests were. Purposeful importations of predators and parasites have occurred also. Most useful insects are usually found in the areas of origin of the citrus species (see Chapter 1).

Although predaceous mites, such as *Euseius hibisci*, and ladybird beetles, *Stethorus picipes*, attack spider mites, the current complexes in California and Florida do not usually control citrus red mite or Texas citrus mite. The same can be said for the citrus rust mite. Populations of both kinds of mites are suppressed in marginal subtropical and tropical climates during rainy periods by entomophagous fungi such as *Hirsutella thompsonii*, but citrus rust mites usually build up to injurious levels in the spring dry period before control by fungi occurs and again in the fall after the summer rains end. In a dry Mediterranean climate, citrus rust mite control was adequate after the introduction and mass rearing of predaceous mites and the judicious use of selective miticides instead of sulfur, which greatly reduces populations of natural predators (Cohen, 1975). A similar experience occurred in California when selective chemicals were integrated with other practices (California IPM Manual Group, 1984).

Citrus thrips are known to scar fruit in late spring during the second generation in California. Orchards with ground cover and minimum spray programmes that avoid organophosphates tend to have fewer thrip problems. Predaceous mites, *Euseius hibisci* and *Anystis agilis*, and the minute pirate bug prey on thrips as do natural enemies found in ground litter. A single spray at petal fall will usually control thrips if other factors are favourable for minimum thrip build-up.

In Florida, a successful biological control programme for blackfly was introduced based on work started in Mexico using two parasitic wasps (*Encarsia opulenta* and *Amitis hesperidum*). These parasites are easily killed by sprays of organophosphates or sulfur, while the blackfly is difficult to kill with insecticides. Best control is maintained by introduction of the parasites and avoidance of these chemicals (Knapp and Browning, 1989). Citrus and woolly whiteflies are controlled under Mediterranean climates by parasitic wasps such as *Cales noacki* and *Eretmocerus* spp. This control is easily disturbed by ants protecting this source of honeydew, dust or insecticide treatments such as sulfur that kill these wasps.

A number of parasites, predators and fungi have been observed associated with aphids in South and Central America including the brown citrus aphid. It is believed that these might control this serious aphid to prevent economic damage. Biological control depends on low levels of a pest to sustain the control species. For an efficient virus vector like the brown citrus aphid, this level of control would probably slow down only slightly the spread of virulent CTV (see next section).

Scale insect pests of citrus are generally under some of the best biological control of any insects (Browning as reported by G. Smith, 1992). Many successful parasitic wasps attacking scale insects are of the genus *Aphytis*. In Florida, purple scale is controlled by *A. lepidosaphes*, a chance introduction, and Florida red scale is controlled by *A. holoxanthus* DeBach, which was purposely introduced (Bullock and Brooks, 1975). In South Africa, the predators *Cilocorus negrita* and *C. distigma* combined with *Aphytis* spp. give good control of Florida red scale. Glover, chaff and citrus snow scale are less well-controlled biologically, but the first two are seldom a problem if repeated sulfur sprays are avoided and some ground cover is maintained to provide a good habitat for predators. Chaff scale is a serious pest in Brazil where clean cultivation diminishes alternate hosts and a higher humidity habitat for predators. Heavy use of sulfur sprays in Brazil kills predators and parasites compounding the problem. *A. lingnanensis* Compere was introduced to Florida for snow scale control, but it has been largely unsuccessful. Some predators (ladybird beetle from China being tested in Australia) and parasites appear to be better biological control agents for this difficult to spray scale insect.

In California, *A. melinus* is an important parasite of the California red scale. Combined with the effects of *Comperiella bifasciata* (Howard), another effective parasite, biological control is obtained. Soft scales like brown soft scale and black scale are parasitized by *Metaphycus* spp. Ladybird beetle larvae and parasitic flies also contribute to soft scale control. The most common cause of upset of these control equilibria is the use of broad spectrum pesticides that kill predators and parasites as well as the target pest. Spray oils and specific target pesticides are essential for effective IPM programmes in controlling scale insects.

The weevil complex is one of the most difficult insect problems currently being researched on the American continent (Futch and McCoy, 1992). Although serious tree damage occurs from Florida to Brazil, the most consistent damage and tree losses occur in the tropical climates of the Caribbean Basin. A number of biological control agents and cultural practice strategies exist that together can give some level of control. Spray programmes have been fairly ineffective as discussed earlier. In the 1940s, it was found that citrus trees could be planted on mounds and then the primary roots exposed so that weevil larvae could not inhabit the area at the base of the trunk to girdle it and the small major roots. This strategy has been combined with the use of a tight mesh black polyethylene screen over the mound. When the larvae fall to the ground from the hatched eggs, they can not penetrate the soil directly under the tree but instead take up residence on lateral feeder root areas at the edge of the canopy. This allows the young tree to become established since water can still penetrate the mesh. Hand labour is often used to collect adults as they feed on the leaves. Both entomophagous fungi, *Beauveria bassiana, Metarrhizium anisopliae* and other species, and *Steinernema* spp., attack the larvae stage in the ground. Predators and parasites also exist but they have not provided consistent control.

Various Lepidoptera attack citrus but usually on an inconsistent basis making it difficult to develop any standardized control practices. A number of predators and parasites exist and can be promoted by sound IPM practices so that these minor pests are even less of a problem.

DISEASES

Worldwide the most serious limitations to profitable production of citrus in otherwise suitable environments are diseases caused by bacteria, mycoplasmas, fungi and viruses. In addition to a number of diseases that cause death or seriously limit production, there are many more problems that require costly control programmes to sustain economic production. Diseases causing fruit drop and blemishes are numerous. A general source of information on citrus diseases is Whiteside *et al.* (1988).

Bacterial Diseases

Some of the most serious diseases of citrus in the world are caused by bacteria. These include citrus canker, greening and citrus variegated chlorosis (CVC). Of much less importance is blast or black pit caused by *Pseudomonas syringae* van Hall. It occurs in cool, wet weather on stems and leaves of grapefruit and oranges (blast) and fruit of lemon (black pit). Both blast and black pit have a stage when a reddish brown coloration occurs after the initial necrosis. Infection occurs through wounds created during wind and heavy rain or similar damaging conditions. In the limited areas where this disease occurs, copper bactericides are applied during the winter or early spring to control the disease (Whiteside *et al.*, 1988).

Citrus canker is a much more serious problem. At least three forms of canker exist (Civerolo, 1981). Asiatic canker (A) is widespread and attacks most citrus species. Most commercial citrus and especially grapefruit are affected. Canker B is primarily a problem on lemons in Argentina, but also occurs in Uruguay. Canker C affects 'Mexican' lime in Brazil. A similar form was thought to occur in Mexico on this cultivar because of the similar lesions on leaves even though no fruit lesions occurred (Garza-Lopez and Medina-Urrutia, 1984), but this disorder is now believed to be an *Alternaria* leaf spot. A fourth form (bacterial leaf spot) has developed in Florida on nursery stock, particularly 'Swingle' citrumelo, but the corky centres of the lesions are not raised. Canker lesions have a characteristic yellow halo around a raised, corky centre. All forms of canker are caused by strains of *Xanthomonas campestris* pathovar *citri*, but they can be distinguished by various immunoassays and genetic markers (Civerolo, 1981).

The bacteria reside on leaf surfaces and can survive for some time, but erupting pustules are the major source of inoculum. The infection of fruit and leaves is usually associated with wind and rain and primarily occurs through injuries. Grapefruit and pummelos are particularly susceptible. Beside injury infections, leaves are also invaded through their stomata from about one-third to two-thirds full leaf expansion. This occurs after stomatal openings are exposed but before the inner stomatal chamber cuticle has developed (Graham *et al.*, 1992). Windblown rain is essential for serious infection. Bacteria-laden water penetrates stomata and injuries to water soak leaves with the bacterial solution. Usually citrus canker does not seriously debilitate the tree, except in some areas of Argentina and Asia, but it does cause fruit losses from abscission and nonmarketable quality due to lesions. In the Maldives (islands about 600 km southwest of India), severe winds and rains apparently lead to serious damage and tree death (Roistacher, 1988). Additionally, fresh citrus fruit from areas with canker are not accepted in citrus areas without the disease.

Several countries such as Japan have stopped producing grapefruit because of its canker susceptibility and the endemic nature of bacterial canker. In non-endemic situations such as the United States, Uruguay and Brazil, eradication has been practised. In Argentina, control by spraying is practised in affected areas. Reasonable control was obtained by three sprays of copper during the spring flush, a time that also covers the most susceptible period of fruit development (Stall *et al.*, 1981). An adverse aspect of this type of programme continued over an extended period is the likely development of copper toxicity (see Chapter 5). Wind breaks are an important element of canker control.

Unlike bacterial canker, greening (yellow shoot in China) and CVC are systemic bacterial diseases of the phloem and xylem, respectively. Symptoms of the two diseases are similar and include leaf chlorosis, production of small upright leaves and some shoot dieback. Greening, caused by a phloem-limited bacterium, occurs throughout Asia and Africa but two distinct types exist (Aubert *et al.*, 1984). The Asian strain is found in warm locales, whereas the African strain exists in cooler climates like northern South Africa. The Asian strain is typically transmitted by the Asian citrus psyllid, *Diaphorina citri* Kuwayama, and the African strain by the African psyllid, *Trioza erytreae* Del Guercio. In some locations like Reunion (East of Madagascar in the Indian Ocean) both forms of greening and both vectors exist in a continuum from low, warm to high, cool locations. The vectors can transmit both forms of the bacterium if they are present. Limes and mandarins may be less affected but all commercial citrus is susceptible. The disease is typically transmitted in the nursery by budding from infected mother trees or by vectors from nearby infected plants shortly after planting. With this early infection, trees never become productive.

In Reunion, biological control has been obtained by introduction of parasites of the vectors and planting of clean stock (Aubert *et al.*, 1984). This programme does not work in areas like southeast Asia where predators and parasites of the vector's parasites exist. In southeast Asia, successful production of citrus has been achieved for nine (Gonzales *et al.*, 1984) and 13 (Indonesia, personal observation) years by eliminating nearby contaminated citrus and citrus relatives, planting clean nursery stock, chemically controlling vector build-up in the spring-summer period and cutting out infected limbs as soon as they show symptoms. A similar programme allows economic production in the northern Transvaal area of South Africa.

A bacterial disease (CVC) has been identified recently in South America and is apparently caused by a strain of the xylem-limited bacterium *Xylella fastidiosa* (Lee *et al.*, 1991). The most characteristic symptoms of this disease are small, hard fruit associated with limbs having yellow mottled leaves and dieback. Chlorotic leaf areas often develop necrotic lesions on the lower surface. This disorder was first identified in the northern part of São Paulo State and adjacent Minas Gerais in Brazil. It can now be found in the central part of São Paulo State's citrus district. Spread by infected budwood is apparent and has been experimentally demonstrated. Vectors are assumed to be sharpshooters since they are known vectors for the Pierce's disease strain in grapes and are present in Brazilian citrus orchards (Lee *et al.*, 1991). The disease has since been reported in northern Argentina and Paraguay.

Mycoplasma-like Diseases

Mycoplasma-like diseases also occur in citrus. These prokaryote-caused diseases are not typical mycoplasma disorders. Stubborn disease is caused by a spiral filamentous organism, *Spiroplasma citri* Saglio *et al.*, that is transmitted by leafhoppers and budding with infected budwood. It can infect most citrus species and is found in California, the eastern Mediterranean, northern Africa and the Middle East. Symptoms include stunting due to short internodes, upright leaves and mottled chlorotic leaf patterns. 'Marsh Seedless' grapefruit, 'Sexton' tangelo and 'Madam Vinous' sweet orange are highly sensitive and used as indicator indexing plants. A second mycoplasma-like disease is witches' broom. It occurs on small-fruited acid limes in Oman and probably in India based on symptoms described by Chadha *et al.* (1970).

Fungal Diseases

Numerous citrus diseases are caused by fungi (Whiteside *et al.*, 1988), but only the most severe will be discussed here. Most of the postharvest fruit decays are fungal-induced and are covered in Chapter 7. A number of the fungal organisms cause fruit blemishes.

The earliest fungal infections during the new fruit season can occur during bloom in warm, humid climates if rains occur. Blossom blight is a slow-growing, orange-pigmented strain of *Colletotrichum gloeosporioides* (Penz.) Sacc. that invades the petals inducing fruitlet abscission (Agostini *et al.*, 1992). The organism is present as appressoria on leaves and as quiescent infections. After wetting, in the presence of dead petals, conidia are produced which infect flowers. About 12–18 h of wetness are required during bloom for the initial process to occur, followed by rain dispersal. This problem is severe throughout tropical America and also occurs in rainy bloom seasons in Brazil, Argentina, and Florida, USA. Benomyl and captafol have provided some control, but in some locations benomyl resistance may have developed. Inducing bloom before the rainy season might avoid some of the problem in tropical areas. Water stress induces flower buds and, where irrigation is available, bloom may be forced by relieving the stress before the rainy season (see bloom induction, Chapter 3).

Sooty blotch, *Gloeodes pomigena* (Schw.) Colby, and sooty mould, *Capnodium citri* Berk. and Desm., are at the other extreme of the spectrum in that the blemish is superficial and does not occur until well into the summer rainy season. Sooty mould grows on the surface and sooty blotch grows and is anchored in the epicuticular wax. These saprophytic fungi mix with dirt. The sooty mould is difficult and sooty blotch is impossible to remove completely during normal packinghouse washing for fresh fruit processing. A summer spray oil will kill and loosen these surface growths. Honeydew from insects contributes to the buildup of these mats, and good insect control is important to minimize these problems at harvest. Chlorine bleaching during bin drenching or washing can reduce the visibility of remaining fungal particles. Severe buildup on leaves can reduce photosynthesis as much as 40% (Wood *et al.*, 1988). Alternaria brown spot of mandarins is caused by *Alternaria citri* Ell. and Pierce. Stem lesions on young flushes are the primary source of inoculum for fruit infections which occur during the first three to four months post bloom. Vigorous, hedged trees produce more flush over a longer time and are more likely to have inoculum for fruit infection than less vigorous and nonhedged trees (Whiteside *et al.*, 1988). 'Dancy' tangerine and 'Minneola' tangelo in Florida and 'Emperor' mandarin in Australia are the primary cultivars affected. Copper sprays during the 3–4 month infection period can reduce the problem, but control is poor in wet seasons and application of excess copper causes toxicity problems. Captafol and iprodione provide effective control. Another strain of this organism causes leaf spotting on rough lemon and 'Rangpur' lime seedlings in nursery tree production, but this is not usually a major problem.

Black spot occurs on lemons and to a lesser extent on other species in South America, southern Africa and Asia. The causal organism, *Guignardia citricarpa* (McAlp.) Petrak., infects the young fruit during the first 4–5 months of growth but the spot symptoms do not appear until near maturity. Both pycnidiospores from mature fruit still on the tree for late maturing cultivars (e.g. Valencia) and ascospores from leaf litter can be sources of inoculum. Between 40 and 180 d are required for mature asci to develop in newly fallen leaves. Spore traps are often used to determine when sprays should be applied since the infection period is so long. Copper sprays are used but up to five applications may be needed. Benomyl was highly effective with one application, but resistance to this chemical may already exist. Avoidance of infected plant material in new plantings is the best way to avoid this problem.

Several fungal disorders affect the fruit and canopy. Melanose is present as a fresh fruit disorder in most of the world where rainy conditions occur during early fruit development. The pycnidia of *Diaporthe citri* Wolf develop in dead wood within about 2 months after the tissue dies. Ascospores produced on older deadwood or leaf litter germinate on young foliage and fruit and penetrate the surface before the cuticle is fully formed, about 3 months after bloom for fruit. Leaf infection is primarily a problem after severe stresses. Freezes in Florida for example, create excessive dead wood and a strong flush is stimulated below the dead wood. Rains or overhead irrigation then wash spores on to the young expanding flush. Grapefruit leaves are often infected even when freeze-damaged wood is not present. Deadwood around hedging cuts serves as a source of inoculum. Fruit infections are evident as small brown raised lesions. In severe cases these lesions can coalesce to form large damaged areas on the fruit, i.e. mudcake melanose on grapefruit, that superficially looks like windscar or rust mite damage. Careful examination with a hand lens will reveal some of the original pustules particularly at the margins of the damaged areas. Control on leaves is not necessary, except that dead wood removal is recommended as soon as practical after a freeze (see Chapter 5). Protectant sprays of copper are used on fruit. In dry springs a single spray 1 month postbloom is sufficient. A second spray may be needed in very wet areas or seasons. In this case the first spray is applied slightly earlier.

Lime anthracnose occurs on 'Key' lime in wet areas. The causal agent, a strain of *Colletotrichum gloeosporioides*, colonizes dead wood and rain splash of spores leads to leaf or fruit infection. Very tender shoots may die back and infection at a later stage causes localized necrosis that dry up and fall out of the leaves (shot holes). Young fruit may abscise, while infection at a later fruit development stage results in corky lesion development. Because of the multiple flushes and fruiting of limes in a single year, protection with copper sprays is not practical. Frequent removal of deadwood may help to reduce disease spread if the disease becomes serious.

Greasy spot caused by *Mycosphaerella citri* Whiteside is perhaps the most serious general cause of yield loss in humid climates in the Americas and Asia. Infected leaves abscise at the end of the first growing season rather than after at least 2 years. This results in a shortage of stored carbohydrates and their redistribution in the tree to stimulate leaf replacement at the expense of fruit production. Alternate year bearing develops with a 25–40% yield reduction in the off-year (Whiteside *et al.*, 1988). Moreover, this alternate bearing can occur even when on-year yields are as little as 15–18 tonnes ha^{-1}.

Infection of new leaves occurs from ascospores released from ascocarps that develop in decomposing leaves during the first part of the rainy season. Germinating spores penetrate through stomata and subsequent hyphal growth in the leaf leads to a yellow mottle on the upper surface followed by development of a raised yellow–brown lesion of the lower surface. At least 2–3 months are required for these symptoms to develop after infection. Location of lesions near the base of leaf blade and main veins appears to accelerate early leaf drop. The peel of grapefruit is also affected after stomatal penetration in the early rainy season. Hyphal development is more restricted in the fruit flavedo and only small dark stomata are evident. The zones of flavedo between these affected stomata do not degreen well in the fall. The overall effect is development of an off-colour green, blotchy area usually on the exposed side of the fruit. In Florida, oil or oil plus copper, if greasy spot was severe the previous year, are applied in July and again later in the summer, if greasy spot has been severe and the summer flush is extensive. In more tropical climates, proper timing of control treatments has not been determined. In Cuba, greasy spot causes severe leaf loss on grapefruit, but very little fruit infection occurs, presumably due to lack of spores when the fruit is susceptible. Attempts to control greasy spot by turning under the leaf litter have not been successful. Removal of leaf litter should provide control if it is done completely, but this practice may not be economically feasible.

At least two biotypes of scab disease are recognized (Whiteside *et al.*, 1984). One attacks rough lemon, 'Temple', sour orange, 'Murcott', grapefruit and will attack sweet orange fruit. The other biotype attacks all citrus except sour and sweet orange and 'Temple'. Citrus scab, *Elsinoe fawcettii* Bitancourt and Jenkins, is widespread and occurs in any citrus-growing area with sufficient postbloom rainfall. Growing areas with Mediterranean-type climates do not have scab. Nursery seedling infections occur on rough lemon, sour orange, 'Rangpur' lime and 'Carrizo' citrange. Shoot apices can be severely distorted. Pustules, hyperplastic host cells under a stroma of mycelia and dead plant cells, form on young fruit after infection by conidia. As little as 2.5 h of continuous wetness promotes the production of conidia. Infection takes an additional 6 h. These processes are not very temperature-sensitive between 16 and 28°C. Leaves are immune by full expansion and fruit reach this stage about 3 months postbloom. Pustule size is positively related to susceptibility, but late infections result in small pustules. Scab and windscar on grapefruit can be confused, but pustules can usually be found at the edges of scarred areas. However, both problems can occur together. Sweet orange scab, *Elsinoe australis* Bitancourt and Jenkins, occurs in South America, but one strain of *E. fawcettii* in Florida does attack sweet orange. Tryon's scab, *Sphaceloma fawcettii* var. *scabiosa* McAlp. and Tyson Jenkins affects lemons in Australia. Control in nurseries is not practical but avoiding overhead irrigation may help and greenhouse production is an alternative. Postbloom sprays of copper can protect the fruit and captafol, benomyl, thiophanate and dithianon are effective where resistance has not developed as is the case for benomyl in Florida (Whiteside *et al.*, 1988). Avoidance of overhead irrigation in the field also reduces disease spread.

Of the fungal diseases primarily affecting the tree, phytophthora gummosis, foot rot and root rot are the most widely spread and major cause of reduced vigour and tree loss. *Phytophthora parasitica* Dast. is the most common species and causes all three symptoms but usually is not found in lesions high in the scaffold. *Phytophthora citrophthora* (R.E. Sm. and E.H. Sm.) Leonian is less common and is more likely to cause gummosis and root rot and gum-producing lesions higher in the scaffold and brown rot on lower fruit. Some infection of elongating root tips occurs by direct penetration by germinating zoospores that are released in free water in the soil. Infection of bark requires injuries. Trunk infections usually girdle and kill the tree, whereas infections of fibrous roots only decrease vigour and productivity. This effect is often detected only by a yield response to treatments that reduce phytophthora levels. The best control is by use of resistant or tolerant rootstocks. Trifoliata and 'Swingle' citrumelo are immune, 'Cleopatra' mandarin, sour orange, most selections of rough lemon and citranges are fairly resistant to foot rot (see Chapter 4). On the other hand, all except *Poncirus trifoliata* and 'Swingle' are susceptible to root rot. Other practices that minimize losses due to phytophthora are budding high for sweet orange scions, use of disease-free seedbeds, nursery fields and potting mixtures and good drainage of orchard sites. For example, many trees are budded from 30 to 45 cm above ground level in lowland tropical areas where phytophthora is a major concern. Flood irrigation, particularly directly on the trunk, should be avoided as should prolonged trunk wetting from direct emitter or microsprinkler irrigation. Moisture absorbing tree wraps and buildup of ant colonies under any kind of tree wrap should be avoided. Ants feed on the trunk and main roots and spores then infect the open wounds. Chemical control using soil-applied metalaxyl or foliar fosetyl-Al is effective. Trunk lesions can be healed with the aid of direct sprays of these fungicides.

Mushroom root rot (armillaria root rot) is not a common problem, but can occur on poorly cleared land where trunks and large roots are not completely removed. Previous stands of oaks, sycamores and eucalyptus are most likely to harbour the causal agents, *Armillaria* spp. Infection usually occurs from healthy root contact with diseased residual root. *Armillaria mellea* (Vahl ex Fr.) Kummer is the most commonly reported causal species in Australia, California and Florida. Complete clearing before planting is the most effective control measure. If disease develops after planting, removal of affected and surrounding citrus with 1 year of fallow before replanting is recommended (Whiteside *et al.*, 1988).

Pink disease caused by *Corticium salmonicolor* Berk. and Br. occurs in tropical, wet growing areas, particularly if copper is not normally applied. There is a wide host range and hyphae probably invade directly. A dense mat of the pink hyphae forms with gumming near the infection mat. Girdling often occurs with necrosis of the limb. Opening the canopy, lifting the tree skirt and pruning out affected limbs followed by copper or lime-sulfur sprays is generally effective to control this problem.

Mal secco occurs in dry, cool climates such as the Mediterranean, Black Sea and Asia Minor citrus-growing areas. It has become the most serious disease of lemons in Italy (Salerno and Cutuli, 1981). The causal agent, *Phoma tracheiphila* (Petri) Kantsch. and Gik., forms pycnidiospores that are carried by rain short distances, or by wind to new leaves where germinating hyphae invade stomata or more likely fresh wounds. The organism also can invade the base of limbs and the trunk through wounds. The xylem is colonized, and this leads to wilt (dry disease). An extended cool, wet period of at least 40 h is required for the sporulation and invasion processes. Excess vigour increases likelihood of infection. Pruning water sprouts before the fall and shallow discing to remove surface roots diminishes infection. Resistant lemon cultivars and rootstocks may provide control in the future (Geraci *et al.*, 1988; De Cicco *et al.*, 1984). Plants were regenerated from tissue cultures that survived exposure to filtrates from *Phoma tracheiphila* and are currently being field tested. Another approach to this problem is generation of somatic hybrids with resistance to mal secco (Tusa *et al.*, 1990). However, this procedure also requires extensive field testing to ensure that horticultural characteristics have not been altered.

Viral Diseases

Historically, viral diseases have been the most destructive diseases of citrus industries worldwide. Citrus tristeza virus (CTV) has been the major problem, but the ringspot–psorosis complex and the viroid complex also are major factors limiting successful citrus production. In addition, several other viruses are of sufficient importance in major production areas to be of concern.

Citrus tristeza virus has been responsible for the loss of millions of trees on sour orange rootstock in Brazil, Argentina, Spain and most recently Venezuela (Mendt *et al.*, 1984). Significant losses have also occurred in California and Florida (USA), South Africa and several other countries. Virulent stains of the disease are endemic in Asia. The virus is transmitted by budding from infected scion trees and by several aphid vectors. The brown citrus aphid (*Toxoptera citricidus* Kirk.) is by far the most efficient vector. It is distributed throughout Asia and South America and is moving rapidly northward through the Caribbean Islands and Central America (Lee *et al.*, 1992).

Based on biological indexing and virulence on different citrus hosts, many virulent and mild strains of CTV exist. Some of the mild strains can effectively cross-protect against some virulent strains (Muller *et al.*, 1984). For example, all citrus nursery trees in South Africa are deliberately inoculated with mild strains of CTV before planting as a means of protecting trees from more severe strains. Some strains cause severe symptoms in one scion species on sour orange but not in others. Moreover, some severe stem-pitting strains affect sweet orange or grapefruit scions directly regardless of the stock on which they are planted. One such strain occurs in Peru and was presumably imported from Asia (Roistacher, 1988). Many of the strains are only transmitted by *T. citricidus* and not the other aphid vectors like *Toxoptera aurantii*, *Aphis gossypii* or *Aphis citricola*. Some of these strains exist in Central American and Caribbean Island citrus even on sour orange. When virulent strains are budded to sour orange, the trees are stunted, but otherwise appear normal in the nursery. The stunting becomes more apparent after 2 or 3 years in the field when fruiting competition for carbohydrates leads to severe stunting and reduced fruit size. Furthermore, a bud union overgrowth may not be noticeable earlier than this time. These trees remain small and unproductive in the orchard.

Most CTV strains do not cause any symptoms in seedling trees, even of sour orange, or other budded combinations on CTV tolerant rootstocks. Stem pitting (pin holes) in the sour stock is the most evident symptom of decline on sour strains but phloem necrosis at the bud union with sour is also fairly specific. Rapid tree decline to death (quick decline) has less specific early symptoms although a pronounced overgrowth at the budunion may occur. Stem pitting of the scion indicates that a very severe CTV strain is present.

Citrus tristeza virus is a flexuous rod-shaped virus, and several immunoassays have been developed for its detection. Polyclonal and monoclonal enzyme-linked immunosorbent assay (ELISA) and dot blot tests exist for general CTV assay and specific identification of some virulent CTV strains (Roistacher, 1992). These techniques rely on antibody binding with the virus, with a colour indicator and to the plate or nitrocellulose membrane. Not all virulent strains have specific antibodies for monoclonal tests and indexing on specific reactive hosts is still required. Seedlings of 'Mexican' lime are used for testing for general virulence (Roistacher, 1992). Genetic mapping of the different strains is being examined as another way to distinguish among strains (Pappu et al., 1993). Genetic engineering to develop resistance by insertion of the virus coat protein gene into the citrus genome is underway. Ability of the plant to produce the virus coat protein inhibits infection by the virus but the mechanism is not understood (Beachy et al., 1990).

Quarantine and disease-free budwood programmes are the preferred current practices for CTV control. In endemic areas, use of nonsymptomatic alternative rootstocks is practised. It is advisable to prevent introduction of new virulent strains and prevent additional spread of existing virulent strains. This is effectively practised only in the South African, Californian and Spanish budwood programmes. Cross-protection with mild strains has been very successful in Brazil for the CTV-sensitive 'Pera' sweet orange (Muller et al., 1984) and in South Africa and Australia for grapefruit.

Tatter leaf and citrange stunt are associated with a rod-shaped viroid. Concave gum results in gum deposits in concentric rings of wood with no external bark symptoms. Chlorotic flecking appears in the leaves. The cause of xyloporosis may be related to this viroid group also.

A virus disorder of satsuma in Japan is caused by a virus that is isometric in shape. Several other disorders including navel orange infectious mottling are apparently caused by this or a related virus.

Leprosis is an important disease in Brazil that occasionally occurs in other South American countries and was reported in Florida. This rhadovirus is transmitted by Brevapalpus phoenicis and perhaps other Brevapalpus spp. It occurs on sweet orange and symptoms on fruit, leaves and twigs start as chlorotic areas that develop into sunken necrotic areas on the fruit and raised necrotic areas on stems and leaves. Bark scaling may also be present. A number of other virus-like, graft-transmissible disorders of minor importance occur on citrus in various parts of the world.

The psorosis–ringspot virus-like complex is widely distributed and has been recognized for some time. Usually trees do not show symptoms until they are 10–20 years old, with bark scaling of the trunk being the most pronounced symptom. The severest strains cause decline of the tree. Ringspot causes gum-impregnated chlorotic spots, blotches and ring spots in leaves. Bark scaling is a typical trunk symptom. Psorosis can produce leaf flecking similar to ringspot. The severe B form of psorosis in Argentina has been associated with the ringspot virus (De Zubrzycki and Zubrzycki, 1984). Psorosis was believed to be only graft-transmitted by budding, but the citrus ringspot-type seems to be spreading in Argentina in presumably psorosis disease-free plantings. Aphids have been suggested as a possible field vector, but alternatively, strains of psorosis may exist that require more stringent methods of detection.

Impietratura has some similarity to ringspot and psorosis, but only young leaf flecking occurs. Gum pockets occur in the fruit with reduced fruit size. It is graft-transmitted and occurs in the Mediterranean basin and southern Africa. Concave gum pockets with oak-leaf patterns is another problem.

The viroids (virus-like also) are responsible for a number of citrus disorders. They consist of much smaller single-stranded RNA molecules than viruses like CTV. Cachexia is most severe on mandarins and some tangelos like 'Orlando'. In addition, some symptom expression occurs in sweet lime, rough lemon and 'Rangpur' lime. Different viroids exist that result in varying degrees of the symptoms which include bumpy trunk bark, bark discoloration and gum deposits. Various degrees of stunting, chlorosis and decline also occur depending on viroid and *Citrus* spp. but the severest combination does not kill the tree. These viroids are slightly smaller than the exocortis viroid. Xyloporosis is part of this complex, but the host range may be slightly different (Whiteside *et al.*, 1988). Presently, it is not clear if different strains of cachexia exist or if there are a number of different viroids that cause similar symptoms. This will not be solved until cloning and regeneration has been done and correlated to biological activity.

Citrus exocortis viroid (CEV) symptom expression occurs primarily on *Poncirus trifoliata* and its hybrids. 'Etrog' citron is also highly reactive and is used for indexing. The more severe strains cause bark scaling, severe tree stunting and reduced yields as typically seen in China and many other countries where budwood is not free of CEV and these rootstocks are used. This disease is graft-transmitted, but no insect vectors are known. Mechanical transmission also occurs, and over a 14-year period this disorder can be completely spread through an orchard by contaminated hedging or pruning tools (Calavan *et al.*, 1981). Spread to mature trees has very little consequence since the major effect is to reduce tree size during canopy development. Tree size control on trifoliata rootstocks has been accomplished by introducing a strain of intermediate severity 1 or 2 years after planting to reduce vigour after some initial size has been obtained (Ashkenazi and Oren, 1988). Many of the mild and moderate strains of CEV are probably other viroids. CEV and other viruses can be eliminated from infected old-line material by correctly applied shoot-tip grafting. This procedure requires removal of the shoot-tip meristem without vascular tissue while it is growing rapidly at high temperature. The shoot-tip is grafted on to a similar size rootstock seedling. Systemic pathogens are not able to invade the fast-growing meristem without vascular paths for rapid movement (Navarro, 1981).

Citrus Blight

A major citrus disease of unknown aetiology, citrus blight is a serious disease occurring in Florida, Brazil, Argentina, Uruguay, Venezuela, South Africa and Queensland, Australia (Timmer et al., 1984). The disease also has been reported in several other tropical countries. It does not occur in countries with cool Mediterranean climates. This disorder has been recognized for over 100 years. Although no causal agent has been found, the disease is readily transmitted by root grafts from infected trees or root pieces to healthy donor trees. In contrast, continuous limb grafts for 6 years did not result in disease transmission (Albrigo et al., 1992), suggesting that the causal agent resides in the root system. Some researchers argue that the cause of citrus blight is abiotic because blight may occur under one soil condition and not in different conditions nearby. It also appears that more blight occurs in high pH soils where lime has been added. These situations probably influence how much symptom development appears, but the clear-cut root transmission in several countries indicates that blight is probably caused by a pathogen.

Symptom development includes specific protein production (Derrick et al., 1990), phloem zinc accumulation associated with a small polypeptide ligand, wood zinc accumulation, then trunk xylem plugging followed by wilt and several secondary nutritional and stress disorders apparently caused by the water stress (Albrigo et al., 1986; Taylor et al., 1988). Trees are not directly killed by citrus blight, but are weakened and usually become infected and killed by *Phytophthora* spp.

Rootstocks vary in susceptibility with rough lemon, 'Volkamer', 'Rangpur' lime and trifoliata being highly susceptible, 'Carrizo' citrange is susceptible, 'Cleopatra' mandarin is moderately susceptible while sour and sweet orange are slightly susceptible. 'Swingle' citrumelo appears to be fairly tolerant (Young et al., 1982). Rootstock choices for growing areas with blight, CTV and phytophthora are very limited (Chapter 4). For example, growers in Venezuela used 'Volkamer' lemon as a major rootstock to replace sour orange rooted trees lost from CTV (Mendt et al., 1984). Unfortunately, these trees are now being lost to citrus blight.

Control Practices

Control of diseases as noted is fairly specific by disease but some general concepts are applicable. Use of disease-free material to avoid many problems is of prime importance. Indexing and clean-up procedures using heat treatment and shoot-tip grafting have been well-reviewed by Navarro (1981) and Roistacher (1988). Procedures for incorporation of these practices into a budwood certification programme are also available (Roistacher, 1992). Choosing rootstocks and scions that are as tolerant or resistant to local diseases as is practical is an important first step in orchard management.

Fungal and bacterial diseases are difficult to control with available pesticides. Copper is used as a protectant for many of these problems, but its use must be balanced against possible development of copper toxicity. Where appropriate, addition of oil or its use alone can reduce the need for multiple applications of copper. Oil also has value in control of several insects (see previous section).

7

FRUIT QUALITY, HARVESTING AND POSTHARVEST TECHNOLOGY

Postharvest handling and processing of citrus fruits are economically important final steps in transferring fruit from the orchard to the consumer. Unlike most other fruit crops, a high percentage of the worldwide citrus crop is processed into frozen concentrated or single strength juice products (Chapter 1). The United States has the largest per capita consumption of orange juice (about 20 l of single strength equivalents annually per person from 1984 to 1986). Western Europe and Canada are two other primary consumers of orange juice. Of interest, most of Asia and eastern Europe consume very small per capita quantities of juice, and remain potentially large untapped markets for citrus juices.

CHARACTERISTICS OF CITRUS FRUIT

As discussed in Chapter 3 (Internal Quality, p. 79), citrus fruit are hesperidium berries. The hesperidium berry differs from other true berries such as tomato and grape in having a leathery peel surrounding the edible portion of the fruit. The fruit peel consists of an outer coloured exocarp (flavedo) and an inner white spongy mesocarp (albedo). The edible portion (endocarp) comprises the interior portion of the carpels which expand into segments containing juice vesicles and seeds (Fig. 7.1). Citrus fruits are unique in having juice vesicles (sacs) emanating from the carpellary membranes. The presence of the leathery rind protects the fruit from damage during handling and desiccation during storage (Albrigo and Carter, 1977).

Citrus fruits are nonclimacteric and thus lack the dramatic rise in ethylene and respiration typical of climacteric fruits such as apple and associated with their becoming edible (ripe). Citrus fruits are also low in starch reserves and thus undergo very slow changes in internal quality during storage. Protracted storage does decrease stored acids converting them to sugars and CO_2 used in respiration. Notable exceptions are lemons (Batchelor

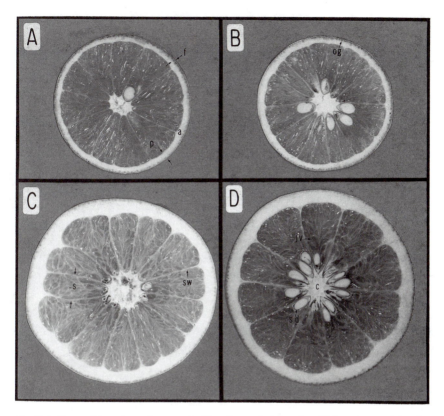

Fig. 7.1. Equatorial sections of orange (A, B) and grapefruit (C, D) and from commercial and seedless (A, C) and seedy (B, D) cultivars. Abbreviations: a, albedo; c, core; f, flavedo; jv, juice vesicle; og, oil gland; p, peel; s, section; sw, section wall; sd, seed. Source: Albrigo and Carter (1977).

and Bitters, 1954) and the 'Palestine' lime (Echeverria and Ismail, 1987), which undergo considerable increases in acid content during curing.

Fruit Composition

The composition of citrus fruits varies with cultivar, climate, rootstock and cultural practices. Most citrus, like other fruits, are primarily water, but also contain over 400 other constituents including moderate levels of carbohydrates, organic acids, amino acids, ascorbic acid and minerals and small quantities of flavonoids, carotenoids, volatiles and lipids. Citrus fruits are low in proteins and fats. Recent work (J. K. Burns, unpublished) suggests that excised juice vesicles may contain less than 1 mg g^{-1} of protein on a fresh

Table 7.1. Composition of California 'Valencia' oranges. The values in the first half of the table are expressed in g (100 g^{-1}) while those in the second half are expressed in mg (100 g)$^{-1}$.

	Peel	Edible portion	Juice
Acid (citric)	0.29	0.75	1.02
Ash	0.78	0.48	0.34
Fat	0.23	0.30	0.29
Moisture	72.52	85.23	87.11
Protein	1.53	1.13	1.00
Sugar			
Reducing	5.56	4.69	4.99
Sucrose	1.99	4.41	4.73
Total	7.55	9.10	9.72
TSS	15.69	13.06	12.59
Ascorbic acid	136.5	39.5	43.5
Biotin	0.005	0.001	trace
Calcium	161.0	36.7	9.5
Carotenoids	9.9	3.4	2.8
Iron	0.8	0.8	0.3
Magnesium	22.2	11.5	11.3
Phosphorus	20.8	21.8	19.5
Potassium	212.0	173.0	163.0
Sodium	3.0	1.3	0.7
Sulfur	21.0	11.5	8.5

Source: Erickson (1968).

weight basis. Extracted juice may have higher levels. Citrus fruits are a good source of pectins and roughage. Much of the increase in orange juice consumption is linked to the potential health-related benefits contained in the juice (Nagy and Attaway, 1980). (Consumers of the 1990s are interested in low fat, high mineral and vitamin C sources of food.)

Total soluble solids

Total soluble solids (TSS), which include carbohydrates, organic acids, proteins, fats and various minerals, comprise from 10 to 20% of the fresh weight of the fruit (Erickson, 1968). Carbohydrates account for 70–80% of the TSS in the fruit. The major groups of carbohydrates in citrus fruits include monosaccharides (glucose, fructose), oligosaccharides (sucrose) and poly-saccharides (cellulose, starch, hemicelluloses, pectins). Sucrose is the primary non-reducing sugar and is the major translocatable carbohydrate (Table 7.1). Fructose and glucose are major reducing sugars and are present at about one-

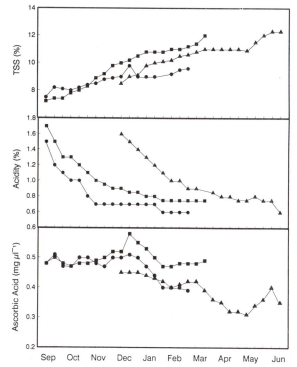

Fig. 7.2. Seasonal changes in juice quality (total soluble solids [TSS], titratable acid [TA] and ascorbic acid) of three sweet orange cultivars. Symbols: ●, 'Hamlin'; ■, 'Pineapple'; ▲, 'Valencia'. Source: Harding *et al.* (1940).

half to equal quantities of sucrose in most citrus juices. Small quantities of mannose and galactose have also been found in citrus juice.

Of the polysaccharides, starch is present in small quantities, particularly as the fruit matures and starch is converted to sucrose, fructose and glucose. Pectin is an important polysaccharide in the cell wall matrix (see following section). Total soluble solids levels increase as fruit size increases, becoming nearly constant or slightly increasing during stage IV of development (Fig. 7.2a).

Organic acids

Total acidity (TA) of citrus juices is an important factor in overall juice quality and in determining time of fruit harvest (Harding *et al.*, 1940). In most citrus-growing regions the ratio of TSS to titratable acids determines whether fruit are harvestable. Organic acids contribute significantly to overall juice acidity. Citric acid is the primary organic acid (70–90% of total) followed by malic and oxalic acids with lesser amounts of succinic, malonic, quinic, lactic, tartaric and other related acids. Sweet limes have very low acid levels which are mostly malic acid (Echeverria and Ismail, 1987). Organic acid levels generally decrease seasonally as citrus fruits mature (Fig. 7.2b). The rate of decrease in acidity is positively correlated with average temperatures during the season. For example, acid levels decrease much more rapidly in low tropical than in subtropical growing regions due to higher mean temperatures (Chapter 3). Higher temperatures increase respiration rates causing less storage of acids in the vacuoles and their more rapid utilization in metabolism. The ratio of TSS:TA increases during maturation and is a good indicator of palatability (see Internal Quality, p. 79).

Ascorbic acid

It has been known for many years that citrus fruits are a valuable source of ascorbic acid (vitamin C). Ascorbic acid functions as a coenzyme and is an essential part of the human diet (Nagy and Attaway, 1980). Levels of ascorbic acid, however, are quite variable among citrus fruits and tend to decrease seasonally (Fig. 7.2c). Ascorbic acid levels are expressed as mg $(100 \text{ ml juice})^{-1}$ and range from 18 to 20 in some tangelos to over 70 for 'Pineapple' sweet oranges. As noted in Table 7.1, ascorbic acid levels are generally higher in the peel than the extracted juice.

Limonin

Triterpene derivatives which produce bitter flavours in the juice are present in most citrus cultivars but affect palatability in only a few cultivars. Limonin was first identified as an important bitter principle in navel oranges in the late 1930s. This substance and other limonoids are present mainly in the peel and are released into the juice at juicing. Navel oranges and 'Honey' mandarins are particularly high in limonoids ranging from 9–19 and 8–25 mg l^{-1}, respectively. Generally values of 6 are considered palatable, 9 as bitter and 24–30 extremely bitter. The range of human detection of limonin in juice is 0.5–32 mg l^{-1} and varies considerably from person to person. Moreover, limonin becomes more difficult to detect as sugar and acid levels increase. Recently developed methods using resin exchange columns are being used commercially to reduce limonoid levels in juice, allowing processors to blend juices typically high in limonin with other juices. Limonin levels decrease seasonally as fruit are held on the tree. In addition, limonin levels vary with rootstock selection. Lowest levels are normally found in fruit from scions on rough lemon rootstock.

Pectins

Pectins are high molecular weight carbohydrates composed of chains of anhydrogalacturonic linkages. They serve as intercellular bonding materials in many fruits and vegetables including citrus. During maturation of citrus fruits, insoluble pectins are converted to water-soluble pectins and pectinates (Nagy and Attaway, 1980). Total pectic substances decrease in the peel and pulp over the season, and water-soluble pectins increase as percentage of the total pectins. This change in pectin composition signals fruit softening or overmaturity. Pectin levels are generally low in the juice. Pectins are used in the manufacture of jellies, jams, preserves, frozen fruits, and as coatings for meats and in baking.

Internal Fruit Quality Standards

Fruit quality standards, which determine minimum levels of palatability and commercial acceptability, have been established empirically over the years for each particular citrus-growing region. The flavour and palatability of citrus fruits is a function of relative levels of TSS, TA and presence or absence of various aromatic or bitter principles. In addition, the juiciness and toughness of the pulp vesicles affects the palatability of fresh citrus eaten out of the hand. However, years of research and observation suggest that the ratio of TSS:TA and in particular the attainment of a certain minimum ratio is a reliable and practical means of assessing fruit quality (Harding and Fisher, 1945). Therefore, in many citrus-growing regions of the world, citrus fruit are deemed to be marketable when a minimum TSS:TA ratio is attained. This minimum ratio varies with location and local standards but generally ranges from 7–9:1 for oranges and mandarins to 5–7:1 for grapefruit. The TSS:TA ratio is not of importance to lemon or lime producers who harvest fresh fruit based on minimum fruit size and juice content and processing fruit on acid and peel oil contents.

Although the TSS:TA ratio is of concern in most regions, the relative levels of TSS (°Brix) also affect palatability. Fruit or juice having high TSS:TA ratios and high Brix taste very sweet, whereas those with low ratios and Brix are tart. Fruit with high ratio and low Brix taste are insipid. Many growing districts use a sliding maturity standard that allows a lower ratio (more acidity) as TSS increase.

Fruit palatability is also a function of culture and tradition. For example, the flavour of sweet oranges is accepted over much of the world, and the rich flavour and deep orange colour of mandarin fruit is prized throughout the world. In contrast, grapefruit are popular in Japan, the United States and northern Europe, whereas the Chinese and Thais prefer the lower acidity and less bitter flavour of pummelos. Therefore, the intended market, fresh or processed, and the particular ethnic flavour preferences influence quality standard levels.

FRESH FRUIT

Production objectives and technologies differ for fresh compared with processing fruit. Although fresh citrus fruit must meet internal quality standards, emphasis is placed on external appearance and fruit size. Consumers in the largest fresh fruit markets demand oranges with bright orange peel coloration, grapefruit with bright yellow or reddish orange coloration, mandarins with orange to red–orange coloration, yellow lemons and green limes. Fruit also must be within certain guidelines for size and must be relatively free of peel blemishes. For example, small grapefruit the size of a mandarin are unmarketable, while conversely overly large mandarins are undesirable in most markets. Moderate-sized grapefruit are in demand in the Japanese market, while large grapefruit are in demand by specialty fruit shippers in the United States market. Similarly, consumers more readily accept yellow vs green lemons, although most fruit are harvested green. The European Community has come to accept only brightly coloured, relatively unblemished fruit that are characteristic of the Mediterranean countries (Chapter 2). Segments of the United States and Central American markets on the contrary will accept fruit with surface blemishes provided that the internal quality is acceptable. Therefore, the most important determinants of citrus fresh fruit quality are fruit size and shape, peel colour and peel quality, i.e. the lack of surface blemishes, peel firmness and texture. Degree of seediness is also important in some markets. In general, commercially seedless fruits (averaging fewer than nine seeds per fruit) (Hensz, 1971) are more desirable than seedy fruits. Perhaps a seed in every other segment is acceptable. The tangerine market in the United States has accepted some seediness, while the European market does not. In these cases, availability of seedless fruit determines market acceptance. Lemon and lime markets accept moderate seediness because fruit are usually sliced and squeezed, although presence of seeds can be annoying when preparing freshly squeezed juice.

Factors Affecting Fruit Size and Shape

Fruit size is a function of several factors including cultivar, rootstock, crop load and cultural practices such as irrigation and nutrition. For example, grapefruit are inherently larger than oranges which are generally larger than mandarins, lemons or limes, although certainly size overlap exists. Consequently, market demands will differ for each species. Most cultivars must be larger than a certain minimum size to be considered marketable. This is especially true for mandarins and mandarin-hybrids that tend to be inherently small and for lemons and limes which are harvested primarily on fruit size and juice content.

Fruit size may be improved through choice of rootstock (see Chapter 4). Generally, lemon-type rootstocks or those which impart vigour to the scion will produce larger fruit than less vigorous rootstocks such as sour orange or *Poncirus trifoliata*. This effect is more pronounced for mandarins and sweet oranges; however, in some instances mandarins become overly large with puffy peels. Fruit with loose or puffy peels ship and store poorly and are more difficult to handle and pack than those with firm peel structure.

Cropload has a significant impact on fruit size especially for mandarins and mandarin-hybrids which tend to alternate bearing. Fruit size in the 'on' year in particular may be unacceptably small for commercial use. There are two methods of regulating cropload in mandarins: the use of growth regulators and pruning. Both methods reduce the cropload in the 'on' year, thus stimulating more crop in the 'off' year and balancing production from year to year. Growth regulators like NAA or ethephon (not registered in some locations) are applied during the second fruit drop period after bloom (physiological drop) when fruit diameter is about 1.5–2 cm. These growth regulators ideally remove about 25–30% of the crop thus allowing the remaining fruits to have higher leaf:fruit ratios and ultimately larger size. The increase in economic returns occurs because fruit below minimum size standards are unmarketable. Hand-pruning or mechanical hedging and topping also removes fruit in the 'on' year and is an effective means of balancing the crop load for many mandarin cultivars. Trees are usually hedged and/or topped before they bloom but after flower bud formation, thus reducing flower number. Newly formed shoots arising from the pruned branches are vegetative but will form flowering shoots in the following season. Alternatively, pruning can be done after the postbloom fruit drop to adjust cropload. This method also reduces leaf area, which is usually replaced by a strong vegetative flush following pruning.

Cultural practices such as irrigation, nutrition and pruning also influence fruit size but may dilute TSS (see the section on TSS in this chapter). Deficient or excessive nitrogen and deficient potassium levels limit fruit size. Hand-pruning is widely used in areas specializing in fresh fruit production such as Spain, Japan and China. Shoots are removed from the centre of the tree and in some cases fruiting shoots are clipped immediately following harvest. Hand-pruning also removes deadwood which is a source of inoculum for various diseases and which may cause limb punctures of the fruit. Hand-pruning improves fruit size, colour and distribution within the tree, especially for mandarins such as satsumas and 'Clementine'.

Variation in shape is usually not of major concern for fresh fruit producers. Fruit shape is not influenced by seed abortion as is the case in apples where misshapen fruit are produced where seeds fail to develop. Fruit shape does vary, however, with climatic conditions. For example, grapefruit under humid subtropical or tropical conditions attain a favourable oblate shape, while those developing under arid conditions become more spherical. Off or late blooms also tend to develop a pronounced neck at the stem end of the fruit termed 'sheepnosing.' Fruit with this characteristic often fail to meet consumer demands. Similarly, mandarins produced under very vigorous growing conditions (lowland tropical) tend to have a puffy peel and often are atypical of the cultivar produced under more moderate climatic conditions.

Factors Affecting Fruit Colour

Under some circumstances postharvest improvement of citrus fruit colour is necessary. From a practical standpoint, attainment of minimum peel colour standards is generally not a problem because fruit may be degreened postharvest using ethylene gas. Fruit must have attained a minimum level of colour, however, before degreening will be successful. For example, oranges developing in low tropical regions cannot be degreened. Degreening of citrus fruit was first observed during rail car shipment of fruit during the 1920s in the United States. Fruit shipped in rail cars containing kerosene heaters were observed to have become less green during shipment. It was subsequently found that acetylene produced by the heaters promoted colour development in the peel. Ethylene, which is structurally similar to acetylene, also induces the degradation of chlorophyll in the peel and in some cases at prolonged cool temperature an enhancement of carotenoid synthesis occurs. Generally, degreening produces the characteristic colour change from green to a lemon yellow in grapefruit and early oranges. After some natural colour development has occurred, the orange undercolour becomes prominent as the chlorophyll is degraded.

From these discoveries, systems have been developed to degreen fruit before sending them through the packing line (Grierson *et al.*, 1978). The exact temperature, relative humidity, ethylene concentrations and temporal conditions for degreening vary with geographic location, cultivar and stage of colour development before degreening. Optimum degreening temperatures range from 25 to 29°C, with considerably less favourable results occurring at temperatures $\leq20°$ or $\geq35°C$. At 15°C chlorophyll degradation is slowed and at 35°C carotenoid synthesis is severely impeded. Generally, relative humidity should be as high as possible, in the range of 90–95% to prevent desiccation and peel breakdown. High humidity also permits healing of minor wounds in the peel that may have developed preharvest or due to rough handling during harvesting. A fungicide drench, usually thiabendazole at 1000 mg l^{-1}, is often used before degreening to reduce decay development (Brown *et al.*, 1988).

Since ethylene gas is a naturally occurring, endogenously produced plant growth regulator, it is effective in degreening at relatively low concentrations between 1 and 10 $\mu l\, l^{-1}$. Durations of degreening vary with stage of fruit development and cultivars. For example, fruits which have undergone colour break degreen more rapidly than those with a dark green peel. Recommended degreening durations vary from 24 to not longer than 72 h. Periods longer than 72 h should be avoided because of desiccation, peel injury development and increased decay. Decay incidence may increase at the longer degreening times even if a fungicide drench is used prior to degreening. Fruit often continue to degreen after being removed from degreening rooms. Degreening conditions, i.e. high humidity and temperature and the presence of ethylene is ideal for development of stem end rot caused by *Diplodia natalensis* (Brown, 1986). Therefore, in regions where this organism is a problem, duration of degreening should be as short as possible to achieve the desired external appearance. *Colletotrichum* (anthracnose) infection in some mandarin cultivars is stimulated by ethylene degreening (Brown and Barmore, 1976).

Peel colour is also enhanced by using dyes in some humid subtropical and tropical growing regions where fruit do not attain peel colour typical of the species. Such is the case with peel colour development of some sweet oranges in Florida. Fruit are flooded in conveyor tanks with food grade red dyes for 1.5–4 min at 46–48°C depending on cultivar. Colour-adding is done only on oranges, 'Temple' oranges and tangelos after degreening.

Factors Affecting Fruit Blemishes

Fruit blemishes are of major concern to fresh fruit growers and marketers. In most regions, fresh fruit grades are based on the extent of peel damage attributable to blemishes. Moreover, considerable time and expense are spent on controlling sources or causal organisms of peel blemishes even though in some cases the internal quality of the fruit is unaffected. In many instances, control is necessary to reduce defoliation and loss of tree vigour. Consumers in most markets consistently indicate that they prefer fruit that are generally free of blemishes with colour typical of the cultivar. As many as 80 different causes for off-grade of citrus are recognized (Albrigo, 1978).

Causes of citrus fruit blemishes

There are numerous biotic and abiotic causes of peel blemishes in citrus, the severity of which varies considerably with growing conditions and geographic location. As stated previously (Chapters 3 and 6), the extent of disease and pest pressures is usually more severe in humid tropical regions than in cool, semiarid or arid subtropical regions. It is impractical to discuss every factor that produces peel blemishes worldwide; however, some of the most widely distributed and economically important types are discussed in the following sections.

Biotic factors

Peel blemishes are caused by a myriad of different insects, mite, fungal, bacterial and other pests (Albrigo, 1978). In some instances the pest merely discolours or disfigures the peel which affects fruit marketability for cosmetic reasons alone. Major insect pests producing blemishes on citrus fruit include scales (soft and armoured), thrips, plant bugs, coffee bean weevil, leaf-miners, crickets, grasshoppers and fruit flies. Armoured scales such as red and purple scale feed on the peel and their armour (carapace) may remain on the fruit even when the insect inside has died. The fruit surface, where the insect is removed by washing during packing, may colour poorly. Thrips produce a distinctive ringing at the stem end of the fruit, but also may cause russeting similar to rust mite damage or scribbling which appears as light brown to grey patches over the entire fruit surface. Plant bugs represent a large group of insects with piercing or sucking mouthparts that puncture the peel causing a wound response, discoloration and sometimes local degreening from ethylene production. Often decay starts in the feeding site and leads to eventual fruit abscission. The coffee bean weevil feeds on fruit and deposits larvae in them. In more tropical climates, leaf-miners feed on fruit, as well as leaves. Crickets and grasshoppers will also feed on immature fruit, leaving large, deep peel depressions. Fruit flies (Mediterranean, Queensland, Caribbean and Mexican) oviposit into the peel. As the larvae develop and feed on the pulp, the fruit softens predisposing it to fungal invasion. Obviously, the development of the larvae within the fruit make it undesirable for consumption. Fruit flies in particular have a severe economic impact on citrus production worldwide. Furthermore, areas having fruit flies cannot ship fruit to fly-free citrus-growing areas unless the fruit have received a cold or hot temperature sterilization treatment (Sharp, 1989) or have been fumigated. However, fumigation now is not allowed in many instances due to environmental and health concerns. Some countries allow shipment from fly-free zones within otherwise infested growing districts. Currently, cold sterilization with pre-conditioning at intermediate temperatures is commonly used (Hatton and Cubbedge, 1982). Experimental evaluation of low dose irradiation is still underway. All methods of fruit fly disinfestation can occasionally result in some fruit injury.

Mites, particularly the citrus rust mite (*Phyllocoptruta oleivora*), are a major source of peel blemishes in many citrus-growing regions. Early feeding injury leads to a sharkskin appearance. In contrast, damage in the summer leads to a russeting, while late damage after complete fruit development results in dark brown, smooth injury (bronzing) (see Chapter 6 for injury process). Extensive rust mite damage causes water loss from the fruit and in some instances fruit drop. The broad mite (*Hemitarsonemus latus*) can cause fruit injury, particularly in warm climates.

Fungi also are a significant cause of peel blemishes. Major fungal diseases include alternaria brown spot (*Alternaria citri*), scab (*Elsinoe fawcettii*), brown rot (*Phytophthora citrophthora*), greasy spot rind blotch (*Mycosphaerella citri*) and melanose (*Diaporthe citri*) (see Chapter 6). Alternaria brown spot is a particular problem for 'Dancy' tangerine and 'Minneola' tangelo. The fungus enters the young fruit peel and produces small black depressions in the peel. These eventually develop into characteristic brown pustules. Scab produces raised lesions on the peel but is a problem primarily on 'Temple' orange, 'Minneola' tangelo, lemon and occasionally grapefruit. Greasy spot rind blotch is characterized by darkening of the stomata and pinkish discoloration of regions surrounding the oil glands of the peel. It is primarily a problem for grapefruit. Melanose in contrast, which is also a problem found primarily on grapefruit, produces small dark, raised lesions on the peel. It is particularly severe where a large amount of deadwood remains in the tree on which the spore-producing pycnidia of the fungus over winter. Brown rot is manifested as firm brown lesions on the peel. Rain or irrigation splash the spores on to the peel where they germinate; therefore, this problem most often occurs on fruit near the ground.

Citrus fruit are also blemished or misshapen by bacteria, namely citrus canker (*Xanthomonas campestris* pv. *citri*), citrus variegated chlorosis (*Xylella fastidiusis* pv. *citri*) and greening (an unnamed fastidious phloem-limited bacterium). Citrus canker is characterized by brownish raised lesions surrounded by a yellow halo. Greening affected fruit become small, irregularly shaped and do not colour properly. Citrus variegated chlorosis, caused by a xylem-limited bacterium, also leads to small misshapen fruit that are overly firm and can damage juice extracting equipment (see Chapter 6).

Citrus fruit are occasionally damaged by birds and small mammals. Bird damage appears as large puncture wounds in the peel differing substantially from much smaller insect-inflicted wounds. Small mammals like rodents gnaw the peel, usually causing fruit drop before the fruit reach the packinghouse.

Abiotic factors

Several abiotic factors are responsible for peel blemishes or defects as well as contributing to internal breakdown of tissues in some instances. These disorders can be further subdivided into those with environmentally related causes such as wind and hail; physiological causes such as creasing, splitting and granulation; mechanical causes such as limb rubs and plugging (rind tearing); or physical and physiological causes combined such as oleocellosis and zebra skin which occur on mandarins (Albrigo, 1978).

Wind is probably the major abiotic factor contributing to peel damage worldwide. Wind causes leaf or limb rubbing abrasions on young fruits before they develop a strong cuticle, until about 2.5 cm diameter. The rubbing kills the outer cell layers of the peel, producing a wound periderm reaction. The peel does not colour properly as a result and develops a light to dark brown blemish. Properly located wind breaks consisting of trees or in some instances man-made structures of shade cloth or similar material serve to reduce the severity of wind scarring. Wind breaks using trees often compete with adjacent citrus trees for nutrients, water and sunlight and may cause tree stunting. Man-made windbreaks are usually cost-prohibitive and are not widely used except with very high cash value cultivars.

Hail damage occurs in a limited number of citrus-growing regions on an irregular basis. Depending on the size of the hail, the peel becomes pitted, sometimes with deeply sunken, darkened areas. The only solution to preventing hail damage is to cover the orchards, a solution which is not cost-effective or practical in most cases, although some covered orchards exist in Japan, primarily for out-of-season mandarin production. Hail causes severe crop losses and even defoliation and limb damage in some growing seasons.

As most physiological disorders of citrus fruit result from a myriad of environmental and physiological conditions, they are difficult to control. Fruit splitting and creasing involve separations of the flavedo or the albedo. Splitting is manifest usually as a longitudinal fissure beginning at the stylar end of the fruit where the peel is thinnest. Splitting is most severe for thin-peeled cultivars including some mandarins such as 'Ellendale' and 'Murcott' as well as sweet oranges like navels, 'Shamouti', 'Hamlin' and some Valencia cultivars. Splitting incidence varies seasonally and is usually greatest where cropload is heavy. It also appears more severe on trifoliate-type than rough lemon-type rootstocks. Splitting may result from water or nutritional stresses early in fruit development, although the fruit splitting occurs later in development. Splitting is more severe at high temperatures and where intensive irrigation is used or heavy rainfall occurs in late season. However, splitting is not strictly related to fruit water relations. For example, amount of rainfall was not correlated with splitting for 'Ellendale' mandarin growing in South Africa (E. Rabe, unpublished). Splitting severity may be decreased by thinning the crop or by a postbloom gibberellic acid (GA_3) spray. Early season spray application of KNO_3 is also effective in reducing splitting in some years.

Creasing involves the separation of the mesocarp (albedo) from the exocarp (flavedo), resulting in sunken areas appearing in the peel. Incidence of creasing also varies seasonally, but appears related to tree and fruit stress early in the season. It is most severe on mandarins and mandarin hybrids and for trifoliate and trifoliate-hybrid rootstocks. Spray applications of GA_3 delay creasing of mandarins and reduce the severity of creasing when applied to navel and 'Valencia' oranges. Creasing also becomes more apparent during storage for 'Robinson' tangerines. An early season K spray can reduce creasing if tree K levels are low but $(NH_4)_3PO_4$ appears to be more beneficial than GA_3 or KNO_3 (Monselise et al., 1976).

Granulation involves the hardening of juice vesicles particularly at the stem and stylar end and late in the harvest season. Mandarins, grapefruit and oranges on vigorous rootstock develop this problem. Granulation, unlike splitting and creasing is not noticeable externally. The problem varies seasonally and is most severe in larger fruit, where a protracted bloom occurs or following heavy summer rains. Fruit with lower than normal TSS and TA as in cultivars on more vigorous rootstocks like rough lemon are more prone to granulation (Erickson, 1968). Grapefruit develop the typical hardened juice vesicles but may have collapsed juice vesicles as well (Hwang et al., 1988). Increases in cell wall development (Burns and Achor, 1989; Hwang et al., 1990) and a higher respiration rate (Burns, 1990) are associated with the hardening of the juice vesicles.

Stylar-end breakdown occurs in 'Tahiti' limes and is associated with excessive fruit turgor. The juice cells in the stylar end of the fruit swell and burst at high temperature with rough handling making the fruit unmarketable. Waiting until fruit have reached a minimum rind oil release pressure (RORP) and thus harvesting later in the morning helps to reduce losses due to stylar end breakdown (Davenport et al., 1976). Blossom end clearing of grapefruit appears to be a similar problem of juice vesicle rupture. Springtime harvest at high temperature and high turgor result in the same problem in the handling of tender grapefruit (E. Echeverria, unpublished). Zebra skin of tangerines behaves similarly but is more related to massive damage of the oil glands on the raised part of the peel.

POSTHARVEST TECHNOLOGY

Harvesting and Handling

Harvesting and handling costs of citrus fruit often equal or exceed total production costs for an orchard. Improper harvesting and handling may cause extensive fruit damage and immediate decreases in packout percentages for fresh fruit and later increases in decay for fresh and processing fruit. Fresh fruit especially must be handled carefully to avoid damage or potential for future decay problems during storage or transit.

Harvesting methods

Most of the world's citrus crop is harvested manually using ladders and some type of picking container. Fruit are either pulled from the limb or clipped in some cases to avoid plugging (rind tearing). In most areas the picker places the fruit into a canvas or plastic sack which he then empties into a larger wooden, plastic or metal container. The container is then moved from the orchard to a truck for transport to the packinghouse or dumped from the containers into large trailers for transport to a processing plant. Fruit are transported out of the orchard by lift trucks, tractor drawn trailers and on

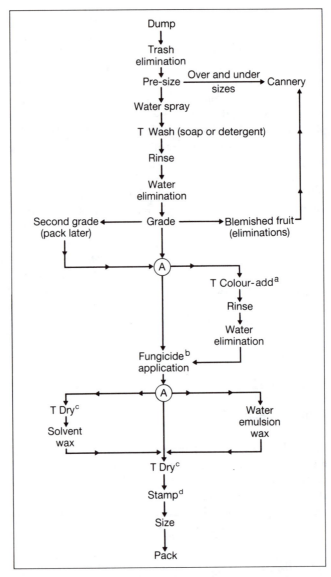

Fig. 7.3. Schematic diagram of a basic packingline layout. Aspects of this design vary with particular local conditions and requirements. Symbols: T, time-dependent operation; A, alternate path. Source: Grierson *et al.* (1978).

[a]Optimal, and only for oranges and tangelos.

[b]Alternative locations for application of fungicides are at the washer, in water emulsion wax or diphenyl pads placed in the packed cartons.

[c]High energy (hot air) drying is prior to solvent wax or following water emulsion wax. Drying following solvent wax is with ambient temperature air.

[d]Stamping can be done prior to a solvent wax, but not prior to a water emulsion wax.

steep slopes on sleds or small vehicles on rail systems (Japan) or manually (China).

Pickers are usually paid on a piece-rate basis but also may work on an hourly rate. There is generally a crew foreman who oversees the harvesting and records the number of containers harvested. The foreman also inspects for improperly harvested fruit, e.g. off size, plugged or fruit harvested with a portion of the limb still attached. The limb may cause puncturing of the adjacent fruit. Plugging (rind tearing) occurs in mandarin type fruit that are not clipped, but poor harvesting leads to plugging of all cultivars. In mandarins, economic analysis would usually indicate that the higher cost for clipping is justified. Plugging predisposes the fruit to fungal infection and desiccation during transit and packing. Foremen are also responsible for ensuring that the entire crop or the properly sized or coloured portion is harvested from a particular orchard or area of an orchard.

Although most citrus cultivars have a rigid, tough peel, poor or delayed handling in the field has a significant deleterious effect on subsequent fruit quality. Reduced fresh marketing is often related to poorer handling, resulting in unreliable delivery condition of the fruit.

Packinghouse Procedures

Packinghouse layout and design vary from basic sorting of fruit by size or colour in the field to high speed ultramodern sorting using a computer. Nevertheless, most packinghouses contain the same basic operations, although certain specializations may be added to adjust for local conditions. The basic packinghouse is outlined in Fig. 7.3. It consists of a bin drencher and degreening rooms (where necessary), an initial wet or dry dumping operation, pregrading, washing, grading, drying, fungicide treating, waxing-drying, sizing, packing and shipping (Grierson *et al.*, 1978).

Degreening with 5 mg l^{-1} ethylene is used in marginal subtropical areas for early harvested cultivars in order to remove the chlorophyll and allow full visualization of the available carotenoid pigments of the peel (see previous section on colour, this chapter).

Directly from the field or following degreening, fruit may be dumped into water that usually contains chlorine and a fungicide (wet dumping) or dumped directly on to a conveyor. In many citrus areas fruit are presorted manually to remove debris and fruit with obvious defects. Dry dumping is usually preferred to keep fungal spores from soaking into wounds. Fruit travel over a conveyor and are pre-wetted, then washed with a mild detergent and water spray rinsed as they rotate on revolving brushes. This washing removes dirt and loosely adhering mould (like sooty mould). High pressure nozzles are occasionally used to remove scale insects. Fruit also may receive a fungistatic treatment of sodium orthophenyl phenate (SOPP) at this time. This material settles in wounds and inhibits fungal growth. At this time, fruit are generally graded to eliminate fungicide and wax treatment of off-grade fruit that will be processed. A fungicide may be applied after this grading stage or it can be incorporated in the wax for application after drying. The fruit are thoroughly dried over absorbent rollers under high velocity, forced air as it is essential to remove water before waxing. If fungicide is applied separately, some dewatering after washing and rinsing is advisable to avoid diluting the fungicide.

During washing, some of the natural wax layers are disturbed causing an increase in weight loss. The natural waxes are apparently not removed by normal washing. Wax is applied to the fruit after drying to prevent this desiccation. Many packinghouses apply natural carnauba or synthetic waxes, which are water soluble, but some use solvent-based synthetic waxes which dry more rapidly than water-based waxes. The wax is applied in a continuous motion via nozzles located above the conveyor or dripped onto the fruit over brushes if a water wax is used. Soft synthetic brushes may be used to polish the fruit before waxing. Wax may include a fungicide at this stage. The wax delays loss of moisture and thereby weight losses during subsequent handling and improves the appearance of the fruit. Fruit are again dried after waxing with heat and forced air. It is preferable that the greatest heat be applied primarily from the wet, entrance side of the dryers. The fruit then continues along the conveyor for final grading.

Most grading worldwide is still done visually. Fruit pass along belts where they are visually evaluated for defects. Usually a certain predetermined percentage of the fruit surface area must be free of blemishes. This percentage varies with grade from <10% to more than 50%. Although this method of grading appears quite subjective, experienced graders are very effective. Alternatively, some packinghouses have high speed photo-grading systems that can detect off-colour or physical defects. Eliminated fruit usually travel along another conveyor to a trailer which transports them to a juice plant. In dry, subtropical areas as much as 95% of the fruit may be packed if of proper size. In humid subtropical and tropical climates, packout may be as low as 50 or 60% because of blemishes and off-colour. In locations like Florida with significant processing capabilities, fruit from low packout orchards are sent directly to a processing outlet.

Selected fruit are then sized for packing. The two conventional methods of sizing are the roller sizer or the use of different sized openings along the conveyor. The roller sizer consists of a conveyor which runs parallel to two tapered rollers which are positioned at wider spacing as the fruit moves from the beginning to end of the roller. Small fruit fall to conveyors and packing areas while larger fruit continue along the rollers until the appropriate spacing between the conveyor and the rollers allow the fruit to drop to other conveyors. The distance between the tapered rollers can be adjusted for each type of citrus to be sized. With the other mechanical method of sizing, fruit are conveyed over an area with different sized openings corresponding to the various size categories. Today, some packinghouses use computer controlled sizers that weigh each fruit individually as it travels in a cup. The computer then directs the cup to dump the fruit into the appropriate size (mass) category along the travel path. Some systems can also size fruit optically.

Fruit are packed into several types of containers but the most common are corrugated cardboard cartons, wooden boxes or netted bags placed in cartons. The standard carton varies from region to region but usually contains between 15 and 18 kg of fruit. Netted shipping bags also vary in size from 1.8 to 6.5 kg. This method often results in having to ship excess weight to meet the minimum net weight marked on the bag (Albrigo *et al.*, 1991). Cartons are usually packed manually or sometimes mechanically using a system of suction cups which lifts a predetermined layer of fruit into the carton placing the fruit into rows. Each carton is then stamped with the number of fruit it contains. This number is inversely related to fruit size. The carton is also stamped with information on types of fungicides used, the cultivar, and name and location of the packer.

Storage, Shipping and Storage Related Disorders

Storage

Citrus fruits are nonclimacteric and have low respiration rates and thus are quite amenable to long-term storage. Nevertheless, storage conditions are cultivar dependent and fruit quality changes occur during prolonged storage. In some areas of the world citrus fruit are held in common storage during winter, or stored in caves where the temperature remains fairly constant all year around (China). Sweet oranges and mandarins in particular may be stored for 2 months or more at 0–4°C with very little loss of fruit quality. Low temperatures depress fruit respiration, water loss and growth of decay organisms. However, lemons, limes and grapefruit cannot be stored at temperatures less than 10°C because they develop chilling injury (CI) which is manifested as pitted necrotic regions in the peel. Consequently, lemons are usually stored at 10–12°C and grapefruit at 10–15°C. The susceptibility to CI varies seasonally, however, and may be related to abscisic acid (ABA), proline, reducing sugar content and water loss of the peel. In Florida, grapefruit harvested before January must be stored at 15°C to avoid CI, whereas fruit harvested after January may be stored at 10°C. Abscisic acid levels tend to decrease in the peel during this period and reducing sugar levels increase. No cause and effect relationships have been established between several of these coincident changes with CI (Purvis, 1981, 1989). Recently, squalene content in the peel has been related to reduced incidence of CI (Nordby and McDonald, 1990). Holding fruit for 1 week at a moderately cool temperature (10–16°C) reduces subsequent injury at temperatures as low as 1°C.

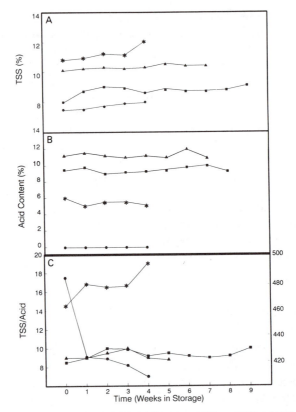

Fig. 7.4. Changes in total soluble solids (TSS) (A), citric acid (B) and TSS:acid (C) for 'Hamlin' orange (■), 'Marsh' grapefruit (▲), 'Robinson' tangerine (✱) and 'Palestine' sweet lime (●) stored at 15°C. Source: Echeverria and Ismail (1987).

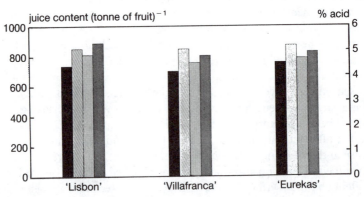

Fig. 7.5. Changes in juice content and acidity of lemons cured for 38 d. Source: Batchelor and Bitters (1954). Symbols: ■, juice (tonne harvest)$^{-1}$; ▨, percentage of acid in harvest; ▨, juice (tonne cured)$^{-1}$; ▨, percentage of acid cured.

Lemon fruit are often harvested when light green and cured during storage at 12.5 to 15°C and a relative humidity (RH) of 95%. Relative humidity should be maintained as high as possible, usually between 85 and 95%, to retard water loss due to vapour pressure gradients between the fruit and air. High humidity also promotes wound healing and growth of some decay organisms such as *Diplodia* stem-end rot.

Controlled atmosphere (CA) storage has modest benefits for citrus fruit storage and is not recommended due to the limited interest and high cost. Citrus fruit are generally not stored for as long as apples, thus common or refrigerated storage are more economical and widely used.

Fruit quality changes during storage and shipment

External and internal changes occur in citrus after it is harvested. Desiccation can lead to unsaleable fruit when 5–10% of weight is lost. The faster the weight loss, the worse the appearance of the peel at a given weight loss. Natural waxes and the postharvest application of wax influence the rate of loss.

Citrus fruit undergo internal quality changes during long-term shipping or storage, but these changes are a function of cultivar and storage conditions. For example, the TSS of 'Marsh' grapefruit did not change during 9 weeks of storage at 15°C, but the TSS of 'Hamlin' orange, increased from 8.1 to 9.5 and 'Robinson' tangerine from 10.9 to 12.1 (Fig. 7.4A). Similarly, citric acid levels decreased for these cultivars but again remained relatively stable for 'Marsh' (Fig. 7.4B). Consequently, TSS:acid ratio increased for 'Hamlin' and 'Robinson' and remained unchanged for 'Marsh' (Fig. 7.4C) (Echeverria and Ismail, 1987). 'Palestine' sweet lime ratios decreased, but the values are so high because of the low acidity that flavour is not affected.

Fruit water content also decreases during storage and shipment. Use of shrinkwrap plastic films on individual fruit is quite effective in reducing water loss, although gas exchange may also be impeded causing anaerobiosis, ethanol and aldehyde accumulation and off-flavours to develop if fruit are held at high temperatures (>30°C) for extended periods. Waxed fruit also develop off-flavours if stored at high temperatures for too long because of poor gas exchange (Hagenmaier and Shaw, 1992). Poorly handled fruit often develop extensive decay in film wraps (Albrigo and Ismail, 1983).

Contrary to behaviour of other citrus, internal quality changes in lemons during curing are quite significant (Fig. 7.5). Juice content increases about 16% primarily from water stored in peel. Acid content also increases significantly from 0.16 to 0.20 mg ml^{-1} (24%) in 4 weeks and peel colour changes from light green to yellow (Batchelor and Bitters, 1954).

Postharvest storage diseases

Fungal diseases are the major source of postharvest fruit damage and losses. Decays usually become evident at the retail and consumer points (Burns and Echeverria, 1988) but are only controlled at the time of harvesting and packing by careful handling and proper use of fungicides. Green (*Penicillium digitatum*) and blue (*Penicillium italicum*) moulds are major sources of fruit decay in the shipping carton. In the case of blue mould, spread of decay in the carton is by contact with other fruit. Eventually most of the carton becomes decayed. Green mould requires open wounds for decay development. Other common postharvest fungal diseases include stem-end decay (*Physalospora rhodina* or *Diaporthe citri*), sour rot (*Endomyces geotrichum*) and brown rot (*Phytophthora citrophthora*). The latter two also spread by contact. These problems can be controlled by using care during picking and handling, proper fungicides and good sanitary measures in the packinghouse to keep damaged, infected fruit at a minimum.

Postharvest nonpathological disorders

Some postharvest disorders are not associated with a pathogen but occur due to handling problems or physiological causes. Oleocellosis occurs when oil cells are ruptured during harvesting or during handling from the field to the packinghouse. The oil from ruptured cells inhibits degreening and discolours the peel, making the fruit unmarketable. Oleocellosis is associated with low rind oil release pressure (RORP) which is a measure of the peel firmness. Lemons and limes are especially susceptible to oleocellosis when harvested below a minimum level of RORP. Therefore, harvest may be delayed until late morning or several hours or days after rainfall until the fruit peel turgor decreases, increasing RORP. Some fruit are also harvested based on differences between the fruit temperature and the wet bulb temperature. This method indirectly determines when the vapour pressure gradient between the fruit and air is sufficient to reduce turgor sufficiently to reduce oleocellosis. The fruit temperature is measured with a thermometer and the wet bulb air temperature with a sling psychrometer. When the fruit temperature exceeds the wet bulb temperature by 2–3°C, harvesting may begin and the incidence of oleocellosis reduced. Careful harvesting and handling also reduces the incidence of oleocellosis. Lemons, which are very susceptible, are often left in the harvesting containers for 12–24 h or overnight before packing. Oleocellosis becomes more visible during grading and can then be removed before packing.

Fruit harvested late in the season and held in long-term storage even under low temperatures are subject to excessive desiccation often causing the peel to become darkened, soft and wrinkled or furrowed around the button. This anomaly is called peel ageing. Stem-end-rind breakdown also occurs near the stem end and is characterized by pitting of the peel except a narrow healthy zone of peel separates the injury from the button. This healthy zone has a heavier wax coating and does not have stomata in its epidermis (Albrigo and Carter, 1977). Film wraps can reduce water loss to 10 or 50% of waxed fruit while improving gas exchange (Ben-Yehoshua et al., 1979), but this method has not replaced waxing primarily because of economic and application constraints (Albrigo and Ismail, 1983).

Citrus Processing

Large quantities of citrus fruit are processed (see Chapter 1). The major product is frozen concentrated orange juice (FCOJ). Other major orange processed products include single strength juice, whole vesicles in juice, squash and canned segments. Grapefruit are processed into the same products. Some mandarins are juiced or canned as segments and lemons and limes are processed into concentrate. Essential oils and pulp are major by-products. The pulp, after drying, is used primarily as cattle feed.

Fruit for processing must be wholesome, having no cuts into juice segment areas or decay, and of suitable quality for product standards. For orange processing, fruit must have TSS, TA and juice content that will produce a high yield and good balance of flavour. Internal orange colour must be sufficient to meet colour score standards or higher coloured juice must be added (blending) to meet these standards (Redd et al., 1986, 1992).

Climate plays an important role in producing high quality juice oranges (Chapter 3). Marginally subtropical climates (e.g. Florida, USA and São Paulo, Brazil) have sufficient cool fall and winter temperatures to stimulate the development of internal orange colour while suppressing respiratory loss of acidity. But the mild fall and winter temperatures allow photosynthesis to continue leading to higher TSS. Fruit also tend to be thin skinned and of high juice content, in excess of 50%.

The most desirable fruit for processing contain >12% TSS, a 14–16 TSS:TA ratio, and a juice colour score of at least 36. Grapefruit from marginally tropical climates are desirable for processing because of their high juice content, low acidity and low levels of bitter components, such as naringen.

Processing

Following harvest, fruit are transported to the processing plant where they are weighed and assigned a lot number. A random fruit sample is then taken to determine the TSS, TA and juice content of the fruit. The percentage juice and TSS determine the kg-solids from which the producer will be paid in ar as such as Florida. Alternatively, many producers are paid simply per tonne of fruit. Many citrus processors also test fruit for wholesomeness, viz. bacterial levels in the juice at this time. Fruit are then either stored for subsequent processing or dumped directly on to conveyors. Some processors manually pregrade fruit removing damaged or unwholesome fruit.

After fruit are washed, juicing is accomplished by sheer press or reamer-type extractors. Reamer-type extractors contain several pairs of cups constructed of stainless steel blades. Fruit are lifted into the lower cup which is meshed with the upper cup hydraulically. As the cups converge the juice and pulp are pressed through a cylindrical strainer that is forced through the centre of the fruit. The remaining peel and rag are ejected for use in cattle feed. The juice, pulp and oils are then transported to the finisher by pipes. The finishers control the size of pulp particles and amount. These machines combine screening and centrifugal force to separate excess large pulp from the juice and to extract essential oils. Concentrating of citrus juices is done using a TASTE evaporator (Terminally Accelerated Short Time Evaporator). Typical evaporator capacities range from 20,000 to 400,000 l of water evaporated h^{-1}. The largest plant has a capacity of over 2,000,000 l h^{-1} of evaporating capacity. This equals about 1000 tonnes of fruit h^{-1}. Juice is usually concentrated to 60–70°Brix for storage and transport.

Storage tank farms are large freezer storage room(s) with 300,000–380,000 l stainless steel tanks connected by stainless steel pipe and pump systems. Smaller blending tanks are also available. Tank farms often have total capacities of 10,000,000 l of concentrate. Some storage occurs in plastic-bag lined 200 l barrels in freezer rooms. The viscous, high Brix juice is transported in these barrels or in bulk in refrigerated truck tankers or bulk tank ships, similar to those used for crude oil transport. Receiving tank farms, particularly for Brazilian orange juice companies, exist in the USA, Europe and Japan (Albrigo and Behr, 1992). Consumer or industrial size packages (containers) are prepared from the bulk juice concentrate out of the tank farms. In many countries, the concentrate is reconstituted to single strength juice for retail sale or blended with other juices or used as a base component for juice-based drinks.

REFERENCES

Agostini, J.P., Timmer, L.W. and Mitchell, D.J. (1992) Morphological and pathological characteristics of strains of *Colletotrichum gloeosporioides* from citrus. *Phytopathology* 82, 1377–1382.

Albrigo, L.G. (1977) Rootstocks affect 'Valencia' orange fruit quality and water balance. *Proceedings of the International Society of Citriculture* 1, 62–65.

Albrigo, L.G. (1978) Occurrence and identification of preharvest fruit blemishes in Florida citrus orchards. *Proceedings of the Florida State Horticultural Society* 91, 78–81.

Albrigo, L.G. and Behr, R.H. (1992) Distribution and consumption consequences for production and marketing of citrus from the American Continent. *Proceedings of the International Society of Citriculture* (in press).

Albrigo, L.G. and Bullock, R.C. (1977) Injury to citrus fruits by leaffooted and citron plant bugs. *Proceedings of the Florida State Horticultural Society* 90, 63–67.

Albrigo, L.G. and Carter, R.D. (1977) Structure of citrus fruits in relation to processing. In: Nagy, S., Shaw, P.E. and Veldhuis, M.K. (eds), *Citrus Science and Technology*. AVI Publishing Co., Connecticut, pp. 33–73.

Albrigo, L.G. and McCoy, C.W. (1974) Characteristic injury by citrus rust mite to orange leaves and fruit. *Proceedings of the Florida State Horticultural Society* 87, 48–55.

Albrigo, L.G. and Ismail, M.A. (1983) Potential and problems of film-wrapping citrus in Florida. *Proceedings of the Florida State Horticultural Society* 96, 329–332.

Albrigo, L.G., Childers, C.C. and Syvertsen, J.P. (1981) Structural damage to citrus leaves from spider mite feeding. *Proceedings of the International Society of Citriculture* 2, 649–652.

Albrigo, L.G., Syvertsen, J.P. and Young, R.H. (1986) Stress symptoms of citrus trees in successive stages of decline due to blight. *Journal of the American Society for Horticultural Science* 111, 465–470.

Albrigo, L.G., McCoy, C.W. and Tucker, D.P.H. (1987) Observations of cultural problems with the 'Sunburst' mandarin. *Proceedings of the Florida State Horticultural Society* 100, 115–118.

Albrigo, L.G., Burns, J.K. and Hunt, F.M. III (1991) Weight loss considerations in

225

preparing and marketing weight-fill bagged citrus. *Proceedings of the Florida State Horticultural Society* 104, 74–77.

Albrigo, L.G., Derrick, K.S., Timmer, L.W., Tucker, D.P.H. and Graham, J.H. (1992) Inability to transmit citrus blight by limb grafts. *Proceedings of the 12th Conference of the International Organization of Citrus Virologists*, pp. 131–138.

Allen, J.C. (1978) The effect of citrus rust mite damage on citrus fruit drop. *Journal of Economic Entomology* 71, 746–750.

Alva, A.K. and Syvertsen, J.P. (1991) Soil and citrus tree nutrition are affected by salinized water. *Proceedings of the Florida State Horticultural Society* 104, 135–138.

Anderson, C.A. and Albrigo, L.G. (1977) Seasonal changes in the relationships between macronutrients in orange (*C. sinensis* Osb.) leaves and soil analytical data in Florida. *Proceedings of the International Society of Citriculture* 1, 20–25.

Anderson, J.A., Gusta, L.V., Buchanan, D.W. and Burke, M.J. (1983) Freezing of water in citrus leaves. *Journal of the American Society of Horticultural Science* 108, 397–400.

Ashkenazi, S. and Oren, Y. (1988) The use of citrus exocortis virus (CEV) for tree size control in Israel – practical aspects. *Proceedings of the International Society of Citriculture* 2, 917–919.

Aubert, B., Sabine, A. and Picard, P. (1984) Epidemiology of the greening disease in Reunion Island before and after the biological control of the African and Asian citrus psyllas. *Proceedings of the International Society of Citriculture* 1, 440–442.

Bailey, L.H. and Bailey, E.Z. (eds) (1978) *Hortus III*. MacMillan, New York.

Bain, J.M. (1958) Morphological, anatomical and physiological changes in the developing fruit of the Valencia orange, *Citrus sinensis* (L.) Osbeck. *Australian Journal of Botany* 6, 1–24.

Barmore, C.R. and Castle, W.S. (1979) Separation of citrus seed from fruit pulp for rootstock propagation using pectolytic enzyme. *HortScience* 14, 526–527.

Barone, E., Bounous, G., Gioffre, D., Inglese, P. and Zappia, R. (1988) Survey and outlook of bergamot (*Citrus aurantium* sub. *bergamia* Sw.) industry in Italy. *Proceedings of the International Society of Citriculture* 4, 1603–1611.

Barrett, H.C. (1985) Hybridization of *Citrus* and related genera. *Fruit Varieties Journal* 39, 11–16.

Barrett, H.C. and Rhodes, A.M. (1976) A numerical taxonomic study of affinity relationships in cultivated *Citrus* and its close relatives. *Systematic Botany* 1, 105–136.

Batchelor, L.D. and Bitters, W.P. (1954) Juice and citric acid content of three lemon varieties. *California Citrograph* 39, 187.

Beachy, R.N., Loesch-Fries, S. and Tumer, N.E. (1990) Coat protein-mediated resistance against virus infection. *Annual Review of Phytopathology* 28, 451–474.

Ben-Yehoshua, S., Kobiler, I. and Shapiro, B. (1979) Some physiological effects of delaying deterioration of citrus fruits by individual seal packaging in high density polyethylene film. *Journal of the American Society for Horticultural Science* 104, 868–872.

Bevington, K.B. and Bacon, P.E. (1977) Effects of rootstocks on the response of navel orange trees to dwarfing inoculations. *Proceedings of the International Society of Citriculture* 2, 567–569.

Bevington, K.B. and Castle, W.S. (1982) Development of the root system of young

'Valencia' orange trees on rough lemon and Carrizo citrange rootstocks. *Proceedings of the Florida State Horticultural Society* 95, 33–37.

Bevington, K.B. and Castle, W.S. (1985) Annual root growth pattern of young citrus trees in relation to shoot growth, soil temperature and soil water content. *Journal of the American Society Horticultural Sciences* 110, 840–845.

Bielorai, H., Dasberg, S., Erner, Y. and Brum, M. (1981) The effect of various soil moisture regimes and fertilizer levels on citrus yield response under partial wetting of the root zone. *Proceedings of the International Society of Citriculture 2*, 585–589.

Bielorai, H., Dasberg, S., Erner, Y. and Brum, M. (1988) The effect of saline irrigation water on Shamouti orange production. *Proceedings of the International Society of Citriculture 2*, 707–715.

Bitters, W.P., Cole, D.H. and McCarty, C.D. (1977) Citrus relatives are not irrelevant as dwarfing stocks or interstocks for citrus. *Proceedings of the International Society of Citriculture 2*, 561–567.

Bitters, W.P., Cole, D.A. and McCarty, C.D. (1981) Effect of budding height on yield and tree size of 'Valencia' orange on two rootstocks. *Proceedings of the International Society of Citriculture 1*, 109–110.

Borrel, M. and Diaz, A. (1981) Effects of mechanical pruning on yield of citrus trees. *Proceedings of the International Society of Citriculture 1*, 190–194.

Boswell, S.B., Lewis, L.N., McCarty, C.D. and Hench, K.W. (1970) Tree spacing of 'Washington' navel orange. *Journal of the American Society for Horticultural Science* 95, 523–528.

Boswell, S.B., McCarty, C.D., Hench, K.W. and Lewis, L.N. (1975) Effect of tree density on the first ten years of growth of 'Valencia' navel orange trees. *Journal of the American Society for Horticultural Science* 100, 370–373.

Bredell, G.S. and Barnard (1977) Microjets for macro-efficiency. *Proceedings of the International Society of Citriculture 1*, 87–92.

Brown, G.E. (1986) Diplodia stem-end rot, a decay of citrus fruit increased by ethylene degreening treatment and its control. *Proceedings of the Florida State Horticultural Society* 99, 105–108.

Brown, G.E. and Barmore, C.R. (1976) The effect of ethylene, fruit colour and fungicides on susceptibility of 'Robinson' tangerines to anthracnose. *Proceedings of the Florida State Horticultural Society* 89, 198–200.

Brown, G.E., Mawk, P. and Craig, J.O. (1988) Pallet treatment with benomyl of citrus fruit on trucks for control of diplodia stem-end rot. *Proceedings of the Florida State Horticultural Society* 101, 187–190.

Browning, H. (1992) CREC report (75 years of excellence in entomology research and extension). *Citrus Industry* 73, 13–14, 36–38.

Buchanan, D.W., Davies, F.S. and Harrison, D.S. (1982) High and low volume under-tree irrigation for citrus cold protection. *Proceedings of the Florida State Horticultural Society* 95, 23–26.

Bullock, R.C. and Brooks, R.F. (1975) Citrus pest control in the USA. In: *Citrus*. Ciba-Geigy Agrochemicals, Basel, Switzerland, pp. 35–37.

Burns, J.K. (1990) Respiratory rates and glycosidase activities of juice vesicles associated with section drying in citrus. *HortScience* 25, 544–546.

Burns, J.K. and Achor, D.S. (1989) Cell wall changes in juice vesicles associated with 'section drying' in stored late-harvested grapefruit. *Journal of the American Society for Horticultural Science* 114, 283–287.

Burns, J.K. and Echeverria, E. (1988) Assessment of quality loss during commercial harvesting and postharvest handling of 'Hamlin' oranges. *Proceedings of the Florida State Horticultural Society* 101, 76–79.

Calavan, E.C., Weathers, L.G., Harjung, M.K. and Blue, R.L. (1981) Spread of exocortis viroid during 14 years in a lemon orchard in southern California. *Proceedings of the International Society of Citriculture* 1, 433–436.

California IPM Manual Group (1984) *Integrated Pest Management for Citrus.* Statewide IPM Project, University of California at Davis, 144 pp.

Camacho-B, S.E. (1981) Citrus culture in the high altitude American tropics. *Proceedings of the International Society of Citriculture* 1, 321–325.

Cameron, J.W. and Soost, R.K. (1979) Sexual and nucellar embryony in F1 hybrids and advanced crosses of *Citrus* with *Poncirus. Journal of the American Society for Horticultural Science* 104, 408–410.

Carpenter, J.B., Burns, R.M. and Sedlacek, R.F. (1981) Phytophthora resistant rootstocks for Lisbon lemons in California. *Citrograph* 67, 287–292.

Cary, P.R. (1981) Citrus tree density and pruning practices for the 21st century. *Proceedings of the International Society of Citriculture* 1, 165–168.

Cassin, J., Bourdeaut, J., Fougue, A., Furan, V., Gaillard, J.P., LeBourdelles, J., Montagut, G. and Moreuil, C. (1968) The influence of climate upon the blooming of citrus in tropical areas. *Proceedings of the International Society of Citriculture* 1, 315–324.

Castle, W.S. (1978) Citrus root systems: their structure, function, growth and relationship to tree performance. *Proceedings of the International Society of Citriculture* 1, 62–69.

Castle, W.S. (1987) Citrus rootstocks. In: Rom, R.C. and Carlson, R.F. (eds), *Rootstocks for Fruit Crops.* John Wiley and Sons, New York, pp. 361–399.

Castle, W.S. and Ferguson, J.J. (1982) Current status of greenhouse and container production of citrus nursery trees in Florida. *Proceedings of the Florida State Horticultural Society* 95, 52–56.

Castle, W.S. and Krezdorn, A.H. (1973) Rootstock effects on root distribution and leaf mineral content of 'Orlando' tangelo trees. *Proceedings of the Florida State Horticultural Society* 86, 80–84.

Castle, W.S. and Rouse, R.E. (1990) Total nutrient content of Florida citrus nursery plants. *Proceedings of the Florida State Horticultural Society* 103, 42–44.

Castle, W.S. and Youtsey, C.O. (1977) Root system characteristics of citrus nursery trees. *Proceedings of the Florida State Horticultural Society* 90, 39–44.

Chadha, K.L., Randhawa, N.S., Bindra, O.S., Chohan, J.S. and Knorr, L.C. (1970) *Citrus Decline in India: Causes and Control.* Punjab Agricultural University, Ohio State University and United States Agency for International Development, India, 97 pp.

Chapman, H.D. (1968) The mineral nutrition of citrus. In: Reuther, W., Batchelor, L.D. and Webber, H.D. (eds), *The Citrus Industry.* University of California Press, California, pp. 127–289.

Chapot, H. (1975) The citrus plant. In: *Citrus,* Technical monograph no. 4, Ciba-Geigy Agrochemicals, Basel, Switzerland, pp. 6–13.

Civerolo, E.L. (1981) Citrus bacterial canker: an overview. *Proceedings of the International Society of Citriculture* 1, 390–394.

Coggins, C.W., Jr. (1981) The influence of exogenous growth regulators on rind quality and internal quality of citrus fruits. *Proceedings of the International Society of Citriculture* 1, 214–216.

Cohen, I. (1975) From biological to integrated control of citrus pests in Israel. In: *Citrus*. Ciba-Giegy Agrochemicals, Basel, Switzerland, pp. 38–41.

Cohen, M. (1968) Exocortis virus as a possible factor in producing dwarf citrus trees. *Proceedings of the International Society of Citriculture* 81, 115–119.

Cohen, M. and Reitz, H.J. (1963) Rootstocks for Valencia orange and Ruby Red grapefruit: results of a trial initiated at Fort Pierce in 1950 on two soil types. *Proceedings of the Florida State Horticultural Society* 76, 29–34.

Crocker, T.E., Bell, W.P. and Bartholic, J.F. (1974) Scholander pressure bomb technique to access the relative leaf water stress of 'Orlando' tangelo scion as influenced by various citrus rootstocks. *HortScience* 9, 453–455.

Dasberg, S.A., Bar-Akiva, A., Spazisky, S. and Cohen, A. (1988) Fertigation vs. broadcasting in an orange grove. *Fertilizer Research* 15, 147–154.

Davenport, T.L. (1990) Citrus flowering. In: Janick, J. (ed.), *Horticultural Reviews*. Timber Press, Portland, Oregon, pp. 349–408.

Davenport, T.L., Campbell, C.W. and Orth, P.G. (1976) Stylar-end breakdown in 'Tahiti' lime – some causes and cures. *Proceedings of the Florida State Horticultural Society* 89, 245–248.

Davies, F.S. (1986a) The navel orange. In: Janick, J. (ed.), *Horticultural Reviews*. AVI Publishing Co., Westport, Connecticut, pp. 129–180.

Davies, F.S. (1986b) Growth regulator improvement of postharvest quality. In: Wardowski, W.F., Nagy, S. and Grierson, W. (eds), *Fresh Citrus Fruits*. AVI Publishing Co., Westport, Connecticut, pp. 79–99.

Davies, F.S. and Maurer, M. (1992) Reclaimed wastewater for irrigation of citrus in Florida. *HortTechnology* 3, 163–167.

De Barreda, D.G. (1977) Present status of weed control practices in Spain. *Proceedings of the International Society of Citriculture* 1, 158–161.

De Cicco, Paradies, V.M. and Salerno, M. (1984) Behaviour of some lemon rootstock towards malsecco root infections: preliminary results. *Proceedings of the International Society of Citriculture* 1, 428–430.

De Zubrzycki, A.D. and Zubrzycki, H.M. (1984) Symptomatology of the disease known as psorosis A in the citrus area of Corrientes, Argentina. *Proceedings of the International Society of Citriculture* 1, 407–409.

Del Rivero, J.M. (1981) Citrus industry in Spain. *Proceedings of the International Society of Citriculture* 2, 973–985.

Derrick, K.S., Lee, R.F., Brlansky, R.H., Timmer, L.W., Hewitt, B.G. and Barthe, G.A. (1990) Proteins associated with citrus blight. *Plant Disease* 74, 168–170.

DeVilliers, J.I. (1969) The effect of differential fertilization on the yield, fruit quality and leaf composition of navel oranges. *Proceedings of the International Society of Citriculture* 1, 1661–1668.

Duncan, L.W. and Cohn, E. (1990) Nematode parasites of citrus. In: Luc, M., Sikora, R.A. and Bridge, J. (eds) *Plant Parasitic Nematodes in Subtropical and Tropical Agriculture*. CAB International, Wallingford, UK.

DuPlessis, S.F. (1984) Crop forecasting for navels in South Africa. *Proceedings of the Florida State Horticultural Society* 96, 40–43.

DuPlessis, S.F. and Koen, T.J. (1988) The effect of N and K fertilization on yield and fruit size of Valencia. *Proceedings of the Sixth International Citrus Congress*, 663–672.

Echeverria, E. and Ismail, M. (1987) Changes in sugars and acids of citrus fruits during storage. *Proceedings of the Florida State Horticulture Society* 100, 50–52.

Economides, C.V. (1976) Performance of Marsh seedless grapefruit on six rootstocks in Cyprus. *Journal of Horticultural Science* 51, 393–400.

Embleton, T.W., Labanauskas, C.K., Jones, W.W. and Cree, C.B. (1963) Interrelations of leaf sampling methods and nutritional status of orange trees and their influence on macro- and micronutrient concentrations in orange leaves. *Proceedings of the American Society for Horticultural Science* 82, 131–141.

Embleton, T.W., Jones, W.W., Pallares, C. and Platt, R.G. (1978) Effects of fertilization of citrus on fruit quality and ground water nitrate-pollution potential. *Proceedings of the International Society of Citriculture* 2, 280–285.

Erickson, L.C. (1968) The general physiology of citrus. In: Reuther, W., Batchelor, L.D. and Webber, H.J. (eds), *The Citrus Industry*. University of California Press, Berkeley, California, pp. 86–126.

Erickson, L.C. and Brannaman, B.L. (1960) Abscission of reproductive structures and leaves of orange trees. *Proceedings of the American Society for Horticultural Science* 75, 222–229.

Erner, Y. and Bravdo, B. (1983) The importance of inflorescence leaves in fruit setting of 'Shamouti' orange. *Acta Horticulturae* 139, 107–113.

Fallahi, E. and Rodney, D.R. (1992) Tree size, yield, fruit quality and leaf mineral nutrient concentration of 'Fairchild' mandarin on six rootstocks. *Journal of the American Society for Horticultural Science* 117, 28–31.

FAO Commodities and Trade Division (1989) Proyecto de Informe. CCP:CI/89 Food and Agriculture Organization of the United Nations, Rome, Italy.

FAO Commodities and Trade Division (1991) Citrus fruit fresh and processed annual statistics. CCP:CI/91 Food and Agriculture Organization of the United Nations, Rome, Italy.

Ferguson, J.J. and Menge, J.A. (1986) Response of citrus seedlings to various field inoculation methods with *Glomus deserticola* in fumigated nursery soils. *Journal of the American Society for Horticultural Science* 111, 288–292.

Florida Agricultural Statistics Department (1990) *Citrus Summary 1989–1990*. Florida Department of Agriculture, Florida, 45 pp.

Ford, H.W. and Tucker, D.P.H. (1975) Blockage of drip irrigation filters and emitters by iron-sulfur-bacterial products. *HortScience* 10, 62–64.

Ford, S.A., Muraro, R.P. and Fairchild, G.F. (1989) Economic comparison of southern and northern citrus production areas in Florida. *Proceedings of the Florida State Horticultural Society* 102, 27–32.

Frost, H.B. and Soost, P.K. (1968) Seed reproduction: development of gametes and embryos. In: Reuther, W., Batchelor, L.D. and Webber, H.J. (eds), *The Citrus Industry*. University of California Press, California, pp. 290–324.

Fucik, J.E. (1977) Hedging and topping in Texas grapefruit orchards. *Proceedings of the International Society of Citriculture* 1, 172–175.

Fucik, J.E. (1978) Sources of variability in sour orange seed germination and seedling growth. *Proceedings of the International Society of Citriculture* 1, 141–143.

Futch, S. and McCoy, C.W. (1992) Citrus root weevils. *Citrus and Vegetable Magazine* 27, 30–33, 36–37.

Gallasch, P.T. and Ainsworth, N.J. (1988) Developments in the Australian citrus industry. *Proceedings of the International Society of Citriculture* 4, 1613–1623.

Garcia-Luis, A., Kauduser, M., Santamarina, P. and Guardiola, J.L. (1992) Low temperature influence on flowering in *Citrus*: the separation of inductive and bud dormancy releasing effects. *Physiologia Plantarum* 86, 648–652.

Gardner, F.E. and Horanic, G.E. (1961) A comparative evaluation of rootstocks for Valencia and Parson Brown oranges on Lakeland find sand. *Proceedings of the Florida State Horticultural Society* 74, 123–127.

Garza-Lopez, J.G. and Medina-Urrutia, V.M. (1984) Diseases of Mexican lime *Citrus aurantifolia* (Christm.) Swingle in Mexico. *Proceedings of the International Society of Citriculture* 1, 311–315.

Geraci, G., Tusa, N. and Somma, V. (1988) Culture filtrates of *Phoma tracheiphila* (Petri) Kanc. et Ghik. to test lemon resistance to mal secco disease. *Proceedings of the International Society of Citriculture*, 829–832.

Gmitter, F.G., Jr. and Hu, X. (1990) The possible role of Yunnan, China, in the origin of contemporary *Citrus* species (*Rutaceae*). *Economic Botany* 44, 267–277.

Gonzales, C.I., Toreja, A.D. and Molino, U.V. (1984) Status of citrus plantings on the ninth year under study for control procedures of greening disease in the Philippines. *Proceedings of the International Society of Citriculture* 1, 409–412.

Graham, J.H. and Syvertsen, J.P. (1984) Influence of vesicular-arbuscular mycorrhizae on the hydraulic conductivity of roots of two citrus rootstocks. *New Phytologist* 97, 277–284.

Graham, J.H. and Syvertsen, J.P. (1985) Host determinants of mycorrhizal dependency of citrus rootstock seedlings. *New Phytologist* 101, 667–676.

Graham, J.H., Gottwald, T.R., Riley, T.D. and Achor, D. (1992) Penetration through leaf stomata and growth of strains of *Xanthomonas campestris* in citrus cultivars varying in susceptibility to bacterial diseases. *Phytopathology* 82, 1319–1325.

Grierson, W., Miller, W.M. and Wardowski, W.F. (1978) *Packingline Machinery for Florida Citrus Packinghouses*. Bulletin 803, University of Florida.

Grimm, G.R. (1956) Preliminary investigations on dieback of young transplanted citrus trees. *Proceedings of the Florida State Horticultural Society* 69, 31–34.

Grosser, J.W. and Gmitter, F.G., Jr. (1990) Protoplast fusion and citrus improvement. In: Janick, J. (ed.), *Plant Breeding Reviews*. Timber Press, Portland, Oregon, pp. 339–374.

Habeck, D.H. (1977) The potential of using insects for biological control of weeds in citrus. *Proceedings of the International Society of Citriculture* 1, 146–148.

Hagenmaier, R.D. and Shaw, P.E. (1992) Gas permeability of fruit coating waxes. *Journal of the American Society for Horticultural Science* 117, 105–109.

Halim, H., Edwards, G.R., Coombe, B.G. and Aspinall, D. (1988) The dormancy of buds of *Citrus sinensis* (L.) Osbeck inserted into rootstock stems: factors intrinsic to the inserted bud. *Annals of Botany* 61, 525–529.

Harding, P.L. and Fisher, D.F. (1945) *Seasonal Changes in Florida Grapefruit*. USDA Technical Bulletin, no. 886, Washington, D.C., 100 pp.

Harding, P.L., Winston, J.R. and Fisher, D.F. (1940) *Seasonal Changes in Florida Oranges*. USDA Technical Bulletin, no. 753, Washington, D.C.

Hare, J.D. and Phillips, P.A. (1992) Economic effects of the citrus red mite (Acari, Tetranychidae) on southern California coastal lemons. *Journal of Economic Entomology* 85, 1926–1932.

Hatton, T.T. and Cubbedge, R.H. (1982) Conditioning Florida grapefruit to reduce chilling injury during low-temperature storage. *Journal of the American Society of Horticultural Science* 107, 57–60.

Hearn, C.J. (1984) Development of seedless orange and grapefruit cultivars through seed irradiation. *Journal of the American Society for Horticultural Science* 109, 270–273.

Hearn, C.J. (1985) Citrus scion improvement program. *Fruit Varieties Journal* 39, 34–37.

Hearn, C.J. (1986) Development of seedless grapefruit cultivars through budwood irradiation. *Journal of the American Society for Horticultural Science* 111, 304–306.

Hearn, C.J. and Hutchison, D.J. (1977) The influence of 'Robinson' and 'Page' citrus hybrids on 10 rootstocks. *Proceedings of the Florida State Horticultural Society* 90, 44–47.

Henrick, C.A. (1982) Juvenile hormone analogs: structure-activity relationship. In: Coat, J.R. (ed.), *Insecticide Mode of Action*. Academic Press, New York, pp. 315–402.

Hensz, R.A. (1971) 'Star Ruby', a new deep-red-fleshed grapefruit variety with distinct tree characteristics. *Journal of the Rio Grande Valley Horticultural Society* 25, 54–58.

Hilgeman, R.H. (1977) Response of citrus trees to water stress in Arizona. *Proceedings of the International Society of Citriculture* 1, 70–74.

Hodgson, R.W. (1967) Horticultural varieties of citrus. In: Reuther, W., Batchelor, L.D. and Webber, H.D. (eds) *The Citrus Industry*. University of California Press, California, pp. 431–591.

Hutchison, D.J. (1974) Swingle citrumelo – a promising rootstock hybrid. *Proceedings of the Florida State Horticultural Society* 87, 89–91.

Hutchison, D.J. (1977) Influence of rootstock on the performance of 'Valencia' sweet orange. *Proceedings of the International Society of Citriculture* 2, 523–525.

Hutchison, D.J. (1985) Rootstock development screening and selection for disease tolerance and horticultural characteristcs. *Fruit Varieties Journal* 39, 21–25.

Hwang, Y.-S., Albrigo, L.G. and Huber, D.J. (1988) Juice vesicle disorders and in-fruit seed germination in grapefruit. *Proceedings of the Florida State Horticultural Society* 101, 161–165.

Hwang, Y.-S., Huber, D.J. and Albrigo, L.G. (1990) Comparison of cell wall components in normal and disordered juice vesicles of grapefruit. *Journal of the American Society of Horticultural Science* 115, 281–287.

Inoue, H. (1990) Effects of temperature on bud dormancy and flower bud differentiation in satsuma mandarin. *Journal of the Japanese Society of Horticultural Science* 58, 919–926.

Ito, M., Ueki, K. and Ito, K. (1981) Approaches to weed management in citrus from the aspect of weed science. *Proceedings of the International Society of Citriculture* 2, 483–485.

Iwagaki, I. (1981) Tree configuration and pruning of satsuma mandarin in Japan. *Proceedings of the International Society of Citriculture* 1, 169–172.

Jackson, L.K. (1991) *Citrus Growing in Florida*. University of Florida Press, Florida.

Jackson, L.K. and Davies, F.S. (1984) Mulches and slow-release fertilizers in a citrus young tree care program. *Proceedings of the Florida State Horticultural Society* 97, 37–39.

Jackson, L.K. and Tucker, D.P.H. (1992) Citrus tree planting and establishment practices. *Citrus Industry* 73, 38–43.

Jahn, O.L. (1973) Inflorescence types and fruiting patterns in 'Hamlin' and 'Valencia' oranges and 'Marsh' grapefruit. *American Journal of Botany* 60, 663–670.

Jahn, O.L. (1979) Penetration of photosynthetically active radiation as a measurement of canopy density of citrus trees. *Journal of the American Society for Horticultural Science* 104, 557–560.

Jones, W.W. and Cree, C.B. (1965) Environmental factors related to fruiting of 'Washington' navel oranges over a 38-year period. *Proceedings of the American Society for Horticultural Science* 86, 267–271.

Jones, W.W. and Embleton, T.W. (1967) Yield and fruit quality of 'Washington' navel orange trees as related to leaf nitrogen and nitrogen fertilization. *Proceedings of the American Society for Horticultural Science* 91, 138–142.

Jordan, L.S. (1981) Weeds affect citrus growth, physiology, yield, fruit quality. *Proceedings of the International Society of Citriculture* 2, 481–483.

Jordan, L.S., Day, B.E. and Jolliffe, V.A. (1969) Residue of herbicides and plant growth regulators in citrus. *Proceedings of the First International Citrus Symposium* 2, 1063–1069.

Kato, T. (1986) Nitrogen metabolism and utilization in citrus. In: Janick, J. (ed.), *Horticultural Reviews*. AVI Publishing Co., Westport, Connecticut, pp. 181–216.

Kato, T., Kubota, S. and Rambang, S. (1982) Uptake and utilization of nitrogen by satsuma mandarin trees in low temperature season. *Bulletin of the Shikoka Agricultural Experimental Station* 40, 1–5.

Kaufmann, M.R. (1977) Citrus – a case study of environmental effects on plant water relations. *Proceedings of the International Society of Citriculture* 1, 57–62.

Kirkpatrick, J.D. and Bitters, W.P. (1968) Physiological and morphological response of various citrus rootstocks to salinity. *Proceedings of the International Society of Citriculture* 1, 391–400.

Kitagawa, H., Matsui, T. and Kawada, K. (1988) Some problems in marketing citrus fruits in Japan. *Proceedings of the International Society of Citriculture* 4, 1581–1587.

Kleinschmidt, G.D. and Gerdeman, J.W. (1972) Stunting of citrus seedlings in fumigated nursery soils related to absence of endomycorrhizae. *Phytopathology* 62, 1447–1453.

Knapp, J.L. (1992) *Florida Citrus Spray Guide*. SP43 Florida Cooperative Extension Service, University of Florida, Florida, 46 pp.

Knapp, J.L. and Browning, H.W. (1989) Citrus blackfly: management in commercial orchards. *The Citrus Industry* 70, 50–51.

Koch, K.E. and Johnson, C.R. (1984) Photosynthate partitioning in split-root citrus seedlings with mycorrhizal and non-mycorrhizal root systems. *Plant Physiology* 75, 26–30.

Koo, R.C.J. (1963) Effects of frequency of irrigation on yield of orange and grapefruit. *Proceedings of the Florida State Horticultural Society* 76, 1–5.

Koo, R.C.J. (1985) Response of 'Marsh' grapefruit trees to drip, undertree sprays and sprinkler irrigation. *Proceedings of the Florida State Horticultural Society* 98, 29–32.

Koo, R.C.J. and Muraro, R.P. (1982) Effect of tree spacing on fruit production and net returns of 'Pineapple' oranges. *Proceedings of the Florida State Horticultural Society* 95, 29–33.

Koo, R.C.J., Anderson, C.A., Calvert, D.A., Stewart, I., Tucker, D.P.H. and Wutscher, H.K. (1984) Recommended fertilizers and nutritional sprays for citrus. Bulletin 536-D, University of Florida Agricultural Experiment Station, Florida.

Kriedmann, P.E. and Barrs, H.D. (1981) Citrus orchards. In: Kozlowski, T.T. (ed.), *Water Deficits and Plant Growth*. Academic Press, New York, pp. 325–417.

Lee, R.F., Derrick, K.S., Baretta, M.J.G., Chagas, C.M. and Rosetti, V. (1991) Citrus variegated chlorosis, a new destructive disease of citrus in Brazil. *The Citrus Industry* 72, 12–13,15.

Lee, R.F., Roistacher, C.N., Niblett, C.L., Rocha-Pena, M., Garnsey, S.M., Yokomi, R.K., Gumpf, D.G. and Dobbs, J.A. (1992) Presence of *Toxoptera citricidus* in Central America, a threat to citrus in Florida and the United States. *The Citrus Industry* 73, 8 pp.

Legaz, F., Ibañez, R., de Barreda, D.G. and Primo Millo, E. (1981) Influence of irrigation and fertilization on productivity of 'Navelate' sweet orange. *Proceedings of the International Society of Citriculture* 2, 591–595.

Leonard, C.D., Stewart, I. and Wander, I.W. (1961) A comparison of ten nitrogen sources for 'Valencia' oranges. *Proceedings of the Florida State Horticultural Society* 74, 79–86.

Lima, J.E.O. (1982) Observations on citrus blight in São Paulo, Brazil. *Proceedings of the Florida State Horticultural Society* 95, 72–75.

Lima, J.E.O. and Davies, F.S. (1984) Growth regulators, fruit drop, yield and quality of navel oranges in Florida. *Journal of the American Society for Horticultural Science* 109, 81–84.

Lo Guidice, V. (1981a) Present status of citrus nematode control in the Mediterranean area. *Proceedings of the International Society of Citriculture* 2, 384–387.

Lo Giudice, V. (1981b) Present status of citrus weed control in Italy. *Proceedings of the International Society of Citriculture* 2, 485–487.

Lord, E.M. and Eckard, M.J. (1985) Shoot development in *Citrus sinensis* L. ('Washington' navel orange). I. Floral and inflorescence ontogeny. *Botanical Gazette* 146, 320–326.

Lord, E.M. and Eckard, M.J. (1987) Shoot development in *Citrus sinensis* L. ('Washington' navel orange). II. Alteration of developmental fate of flowering shoots after GA$_3$ treatment. *Botanical Gazette* 148, 17–22.

Lovatt, C.J., Streeter, S.M., Minter, T.C., O'Connell, N.V., Flaherty, D.L., Freeman, M.W. and Goodall, P.B. (1984) Phenology of flowering in *Citrus sinensis* (L.) Osbeck, cv. 'Washington' navel orange. *Proceedings of the International Society of Citriculture* 1, 186–190.

Lovatt, C.J., Zheng, Y. and Hake, K.D. (1988) Demonstration of a change in nitrogen metabolism influencing flower initiation in *Citrus*. *Israel Journal of Botany* 37, 181–188.

Matthews, G.A. (1982) *Pesticide Application Methods*. Longman: Harlow, UK, 336 pp.

Marler, T.E. and Davies, F.S. (1987) Growth of barerooted and container-grown 'Hamlin' orange trees in the field. *Proceedings of the Florida State Horticultural Society* 100, 89–93.

Marler, T.E. and Davies, F.S. (1988) Soil water content and leaf gas exchange of young field-grown 'Hamlin' orange trees. *Proceedings of the Interamerican Society of Tropical Horticulture* 32, 51–64.

Marler, T.E. and Davies, F.S. (1990) Microsprinkler irrigation and growth of young

'Hamlin' orange trees. *Journal of the American Society for Horticultural Sciences* 115, 45–51.

Marler, T.E., Ferguson, J.J. and Davies, F.S. (1987) Growth of young 'Hamlin' orange trees using standard and controlled-release fertilizers. *Proceedings of the Florida State Horticultural Society* 100, 61–64.

Maust, B. (1992) Nutrition of citrus trees in the nursery. MSc thesis, University of Florida, Gainesville.

Maxwell, N.P. and Rouse, R.E. (1984) Growth and yield comparison of ten-year-old red grapefruit trees from field- and container-grown nursery stock. *Journal of the Rio Grande Valley Horticultural Society* 37, 71–73.

McCoy, C.W. (1985) Current status of biological control in Florida. In: Hoy, M.A. and Herzog, D.C. (eds), *Biological Control in Agricultural IPM Systems*. Academic Press, New York, pp. 481–499.

McCoy, C.W. (1988) The biology of the citrus rust mite and its effects on fruit quality. In: Ferguson, J.J. and Wardowski, W.F. (eds), *Citrus Short Course Proceedings – Factors Affecting Fruit Quality*. Florida Cooperative Extension Service, Florida, pp. 54–68.

McCoy, C.W. and Albrigo, L.G. (1975) Feeding behavior and nature of injury to the orange caused by citrus rust mite. *Phyllocoptruta oleivora* (Prostigmata, Eriophyoidea). *Annals of the Entomological Society of America* 68, 289–297.

Mendel, K. (1969) The influence of temperature and light on the vegetative development of citrus trees. *Proceedings of the First International Citrus Symposium* 1, 259–265.

Mendt, R., Boscan, R., Sabogal, L., Perez, G., Plaza, G. and Lastra, R. (1984) Spread and effects of citrus tristeza virus in Venezuela. *Proceedings of the International Society of Citriculture* 1, 435–437.

Menge, J.A., Lembright, H. and Johnson, E.L.V. (1977) Utilization of mycorrhizal fungi in citrus nurseries. *Proceedings of the International Society of Citriculture* 1, 129–132.

Monselise, S.P. and Halevy, A.H. (1964) Chemical inhibition and promotion of citrus flower bud induction. *Proceedings of the American Society for Horticultural Sciences* 84, 141–146.

Monselise, S.P., Weiser, M., Shafir, N., Goren, R. and Goldschmidt, E.E. (1976) Creasing of orange peel – physiology and control. *Journal of Horticultural Science* 51, 341–351.

Moss, G.I. (1978) Propagation of citrus for future plantings. *Proceedings of the International Society of Citriculture* 1, 132–135.

Muller, G.W., Sobrinho, J.T., Pompeu, J. Jr. and Costa, A.S. (1984) Cross protection increases yield of tristeza tolerant sweet orange. *Proceedings of the International Society of Citriculture* 1, 349–350.

Mungomery, W.V., Jorgensen, K.R. and Barnes, J.A. (1978) Rate and timing of nitrogen application to navel oranges: effects on yield and fruit quality. *Proceedings of the International Society of Citriculture* 1, 285–288.

Nagy, S. and Attaway, J.A. (eds) (1980) *Citrus Nutrition and Quality*. American Chemical Society, Washington, D.C.

Navarro, L. (1981) Citrus shoot-tip grafting *in vitro* (STG) and its application: a review. *Proceedings of the International Society of Citriculture* 1, 452–456.

Nemec, S. (1978) Response of six citrus rootstocks to three species of *Glomus*, a mycorrhizal fungus. *Proceedings of the Florida State Horticultural Society* 91, 10–14.

Newcomb, D.A. (1977) Citrus seed production. *Proceedings of the International Society of Citriculture* 1, 124–126.

Nordby, H.E. and McDonald, R.E. (1990) Squalene in grapefruit wax as a possible natural protectant against chilling. *Lipids* 25, 807–810.

O'Bannon, J.H. and Ford, H.W. (1977) Resistance in citrus rootstocks to *Radopholus similis* and *Tylenchulus semipenetrans* (Nematoda). *Proceedings of the International Society of Citriculture* 2, 544–549.

O'Bannon, J.H., Chew, V. and Tomerlin, A.T. (1977) Comparison of five populations of *Tylenchulus semipenetrans* to *Citrus*, *Poncirus* and their hybrids. *Journal of Nematology* 9, 162–165.

Oberholzer, P.C.J. (1969) Citrus production in southern Africa. *Proceedings of the First International Citrus Symposium* 1, 111–120.

Oren, Y. and Israeli, E. (1977) Herbicide application through irrigation systems (herbigation) in citrus. *Proceedings of the International Society of Citriculture* 1, 152–154.

Oslund, C.R. and Davenport, T.L. (1987) Seasonal enhancement of flower development in 'Tahiti' limes by marcottage. *HortScience* 22, 498–501.

Pappu, H.R., Pappu, S.S., Niblett, C.L., Lee, R.F. and Civerolo, E. (1993) Comparative sequence analysis of coat proteins of biologically distinct citrus tristeza closterovirus isolates. *Virus Genes* 7, 255–264.

Parsons, L.R., Wheaton, T.A., Faryna, N.D. and Jackson, J.L. (1991) Improve citrus freeze protection with elevated microsprinklers. *Proceedings of the Florida State Horticultural Society* 104, 144–147.

Passos, D.S., Cunha, A.P., Coelho, Y.S. and Rodriques, E.M. (1977) Behavior of orange trees under three spacings in the state of Bahia, Brazil. *Proceedings of the International Society of Citriculture* 1, 169–171.

Phillips, R.L. (1974) Performance of 'Pineapple' orange trees at three spacings. *Proceedings of the Florida State Horticultural Society* 87, 81–84.

Phillips, R.L. (1980) Hedging and topping practices for Florida citrus. *Citrus Industry* 61, 5–10.

Phung, H.T. and Knipling, E.B. (1976) Photosynthesis and transpiration of citrus seedlings under flooded conditions. *HortScience* 11, 131–133.

Purvis, A.C. (1981) Free proline in peel of grapefruit and resistance to chilling injury during cold storage. *HortScience* 6, 160–161.

Purvis, A.C. (1989) Soluble sugars and respiration of flavedo tissue of grapefruit stored at low temperatures. *HortScience* 24, 320–322.

Rabe, E. (1991) Bench-rooted citrus nursery trees. *The Citrus Industry* 72, 52–53.

Rabe, E. and van der Walt, H.P. (1992) Effect of urea sprays, girdling and paclobutrazol on yield in 'Shamouti' sweet orange trees in a subtropical climate. *Journal of the South African Society of Horticultural Sciences* 2, 77–81.

Raghuvanshi, S.S. (1968) Cytological evidence bearing on evolution in citrus. *Proceedings of the First International Citrus Symposium* 1, 207–214.

Rasmussen, G.F. and Smith, P.F. (1961) Evaluation of fertilizer practices for young orange trees. *Proceedings of the Florida State Horticultural Society* 74, 90–95.

Redd, J.B., Hendrix, C.M., Jr. and Hendrix, D.L. (1986) *Quality Control Manual for Citrus Processing Plants*, Vol. I. Intercit Incorporated, Florida.

Redd, J.B., Hendrix, D.L. and Hendrix, C.M., Jr. (1992) *Quality Control Manual for Citrus*

Processing Plants, Vol. II. AgScience Incorporated, Florida.

Reuther, W. and Rios-Castano, D. (1969) Comparison of growth, maturation and composition of citrus fruits in subtropical California and tropical Colombia. *Proceedings of the First International Citrus Symposium* 1, 277–300.

Rieger, M. (1989) Freeze protection of horticultural crops. In: Janick, J. (ed.), *Horticultural Reviews*, Vol. XI. Timber Press, Oregon, pp. 46–109.

Rodriquez, O. and Moreira, S. (1969) Citrus nutrition – 20 years of experimental results in the state of São Paulo, Brazil. *Proceedings of the First International Citrus Symposium* 3, 1579–1586.

Rogers, J.S. and Bartholic, J.F. (1976) Estimated evapotranspiration and irrigation requirements for citrus. *Proceedings of the Soil and Crop Science Society of Florida* 35, 111–117.

Roistacher, C.N. (1988) Concepts in the detection and control of citrus virus and virus-like diseases. *Proceedings of the International Society of Citriculture* 2, 853–862.

Roistacher, C.N. (1992) *Graft-transmissible Diseases of Citrus – Handbook for Detection and Diagnosis.* FAO, Italy, 286 pp.

Roose, M.L. (1988) Isozymes and DNA restriction fragment length polymorphisms in *Citrus* breeding and systematics. *Proceedings of the International Society of Citriculture* 1, 57–67.

Roth, R.L., Rodney, D.R. and Gardner, B.R. (1974) Comparison of irrigation methods, rootstocks and fertilizer elements on 'Valencia' orange trees. *Proceedings of the Second International Drip Irrigation Congress* 2, 103–108.

Rouse, R.E. (1982) Evaluation of 5 media and 5 fertilizer treatments for container citrus. *Journal of the Rio Grande Valley Horticultural Society* 35, 167–171.

Rouse, R.E. (1988) Bud forcing method affects budbreak and scion growth of citrus grown in containers. *Journal of the Rio Grande Valley Horticultural Society* 41, 69–73.

Russo, F. (1981) Present situation and future prospect of the citrus industry in Italy. *Proceedings of the International Society of Citriculture* 2, 969–973.

Salerno, M. and Cutuli, G. (1981) The management of fungal and bacterial diseases of citrus in Italy. *Proceedings of the International Society of Citriculture* 1, 360–362.

Saunt, J. (1990) *Citrus Varieties of the World.* Sinclair International, UK.

Savage, E.M. and Gardner, F.E. (1965) The origin and history of Troyer and Carrizo citranges. *The Citrus Industry* 46, 4–7.

Schneider, H., Platt, R.G. and Bitters, W.P. (1978) Diseases and incompatibilities that cause decline in lemons. *Citrograph* 63, 219–221.

Scora, R.W. (1988) Biochemistry, taxonomy and evolution of modern cultivated *Citrus. Proceedings of the International Society of Citriculture* 1, 277–289.

Scora, R.W. and Kumamoto, J.J. (1983) Chemotaxonomy of the genus *Citrus*. In: Waterman, P.G. and Grundon, M.F. (eds), *Chemistry and Chemical Taxonomy of the Rutales.* Academic Press, London, pp. 343–351.

Sharp, J.L. (1989) Preliminary investigation using hot air to disinfect grapefruit of Caribbean fruit fly immatures. *Proceedings of the Florida State Horticultural Society* 102, 157–159.

Sharples, G.C. and Hilgeman, R.H. (1969) Influence of differential nitrogen fertilization on production, tree growth, fruit size and quality, and foliage composition of 'Valencia' orange trees in central Arizona. *Proceedings of the First International Citrus Symposium* 3, 1569–1578.

Singh, M., Tamma, R.V. and Nigg, H.N. (1989) HPLC identification of allelopathic compounds from *Lantana camara. Journal of Chemical Ecology* 15, 81–89.

Smajstrla, A.G. and Koo, R.C.J. (1984) Effects of trickle irrigation methods and amounts of water applied on citrus yields. *Proceedings of the Florida State Horticultural Society* 97, 3–7.

Smajstrla, A.G., Parsons, L.R., Aribi, K. and Velledis, G. (1985) Responses of young citrus trees to irrigation. *Proceedings of the Florida State Horticultural Society* 98, 25–28.

Smith, P.F. (1966a) Citrus nutrition. In: Childers, N.F. (ed.), *Fruit Nutrition.* Horticultural Publications, New Jersey, pp. 174–207.

Smith, P.F. (1966b) Leaf analysis of citrus. In: Childers, N.F. (ed.), *Fruit Nutrition.* Horticultural Publications, New Jersey, pp. 208–228.

Smith, P.F. (1969) Effects of nitrogen rates and timing of application on 'Marsh' grapefruit in Florida. *Proceedings of the First International Citrus Symposium* 3, 1559–1567.

Smith, P.F. and Reuther, W. (1953) Mineral content of oranges in relation to fruit age and some fertilization practices. *Proceedings of the Florida State Horticultural Society* 66, 80–85.

Snowball, A.N., Halligan, E.A. and Mullins, M.G. (1988) Studies on juvenility in *Citrus. Proceedings of the International Society Citriculture* 1, 467–473.

Southwick, S.M. and Davenport, T.L. (1986) Characterization of water stress and low temperature effects on floral induction in citrus. *Plant Physiology* 81, 26–29.

Spurling, M.B. (1969) Citrus in the Pacific area. *Proceedings of the First International Citrus Symposium* 1, 93–101.

Stall, R.E., Miller, J.W., Marco, G.M. and Canteros de Echenique, B.I.C. (1981) Timing of sprays to control cancrosis of grapefruit in Argentina. *Proceedings of the International Society of Citriculture* 1, 414–417.

Stansly, P.A., Rouse, R.E., McGovern, R.J. and Davenport, S.B. (1991) Chemical deterrents to girdling of young citrus by subterranean termites. *Proceedings of the Florida State Horticultural Society* 104, 156–159.

Suzuki, K. (1981) Weeds in citrus orchards and their control in Japan. *Proceedings of the International Society of Citriculture* 2, 489–492.

Swietlik, D. (1992) Yield, growth and mineral nutrition of young 'Ray Ruby' grapefruit trees under trickle or flood irrigation and various nitrogen rates. *Journal of the American Society for Horticultural Sciences* 117, 22–27.

Swingle, W.T. (1948) Botany of citrus and its wild relatives of the orange subfamily. In: Webber, H.J. and Batchelor, L.D. (eds), *The Citrus Industry.* University of California Press, California, pp. 129–174.

Swingle, W.T. and Reece, P.C. (1967) The botany of citrus and its wild relatives. In: Reuther, W., Batchelor, L.D. and Webber, H.J. (eds), *The Citrus Industry.* University of California Press, California, pp. 190–340.

Syvertsen, J.P. (1981) Hydraulic conductivity of four commercial citrus rootstocks. *Journal of the American Society for Horticultural Science* 106, 378–381.

Syvertsen, J.P. (1982) Minimum leaf water potential and stomatal closure in citrus leaves of different ages. *Annals of Botany* 49, 827–834.

Syvertsen, J.P. (1984) Light acclimation in citrus leaves. II. CO_2 assimilation and light, water and nitrogen use efficiency. *Journal of the American Society for Horticultural Sciences* 109, 812–817.

Syvertsen, J.P. and Graham, J.H. (1985) Hydraulic conductivity of roots, mineral nutrition and leaf gas exchange of citrus rootstocks. *Journal of the American Society for Horticultural Science* 110, 865–869.

Syvertsen, J.P., Zablotowicz, R.M. and Smith, M.L. (1983) Soil temperature and flooding effects on two species of citrus. I. Plant growth and hydraulic conductivity. *Plant and Soil* 72, 3–12.

Tachibana, S. and Nakai, S. (1989) Relation between yield and leaf area index in different planting densities under different cultural treatments in 'Satsuma' mandarin (*Citrus unshiu* Marc. var. *praecox*) tree. *Journal of the Japanese Society of Horticultural Science* 57, 561–567.

Talhouk, A.S. (1975) Citrus pests throughout the world. In: *Citrus*. Ciba-Geigy Agrochemicals, Basel, Switzerland, pp. 21–23.

Tanaka, T. (1977) Fundamental discussion of *Citrus* classification. *Studia Citrologica* 14, 1–6.

Taylor, K.C., Albrigo, L.G. and Chase, C.D. (1988) Zinc complexation in the phloem of blight-affected citrus. *Journal of the American Society for Horticultural Science* 113, 407–411.

Timmer, L.W., Brlansky, R.H., Lee, R.F., Graham, J.H., Agostini, J.P., Fischer, H.U. and Casafus, C. (1984) Characteristics of citrus trees affected with blight in Florida, by declinamiento in Argentina and by declinio in Brazil. *Proceedings of the International Society of Citriculture* 1, 371–374.

Tolkowsky, S. (1938) *Hesperides*: a history of the culture and use of citrus fruits. John Bales, Sons & Curnow, 371 pp.

Torres, A.M., Soost, R.K. and Mau-Lastovicka, T. (1978) Citrus isozymes. *Journal of Heredity* 73, 335–339.

Tucker, D.P.H. and Singh, M. (1992) Weeds. In: Knapp, J.L. (ed.), *Florida Citrus Spray Guide* (SP 43). University of Florida, Gainesville, pp. 16–23.

Turrell, F.M. (1961) Growth of the photosynthetic area of *Citrus*. *Botanical Gazette* 122, 285–298.

Tusa, N, Grosser, J.W. and Gmitter, F.G. (1990) Plant regeneration of 'Valencia' sweet orange, 'Femminello' lemon and the interspecific somatic hybrid following protoplast fusion. *Journal of the American Society for Horticultural Science* 115, 1043–1046.

Van Bavel, C.H., Newman, M. and Hilgeman, R.H. (1967) Climate and estimated water use by an orange orchard. *Agricultural Meteorology* 4, 27–37.

Vandiver, V.V. (1992a) Ditch bank, emerged and floating weeds. In: Knapp, J.L. (ed.), *Florida Citrus Spray Guide* (SP 43). University of Florida, Gainesville, pp. 32–42.

Vandiver, V.V. (1992b) Submerged aquatic weeds. In: Knapp, J.L. (ed.), *Florida Citrus Spray Guide* (SP 43). University of Florida, Gainesville, pp. 24–31.

Von Staden, P.F.A. and Oberholzer, P.C.J. (1977) The performance of nucellar citrus lines on several rootstocks in South Africa. *Proceedings of the International Society of Citriculture* 2, 532–534.

Webber, J.H., Reuther, W. and Lawton, H.W. (1967) History and development of the citrus industry. In: Reuther, W., Webber, H.J. and Batchelor, L.D. (eds), *The Citrus Industry*. University of California Press, Riverside, pp. 1–39.

Wen-cai, Z. (1981) Development and outlook of citrus industry in China. *Proceedings of the International Society of Citriculture* 2, 987–990.

Wethern, M. (1991) Citrus debittering with ultracentrifugation/absorption combined technology. *Transactions of the Citrus Engineering Conference*, American Society of Agricultural Engineers, pp. 48–66.

Wheaton, T.A. (1981) Fruit thinning of Florida mandarin using plant growth regulators. *Proceedings of the International Society of Citriculture* 1, 263–268.

Wheaton, T.A., Castle, W.S., Tucker, D.P.H. and Whitney, J.P. (1978) Higher density plantings for Florida citrus: concepts. *Proceedings of the Florida State Horticultural Society* 91, 27–33.

Wheaton, T.A., Castle, W.S., Whitney, J.D., Tucker, D.P.H. and Muraro, R.P. (1990) A high density citrus planting. *Proceedings of the Florida State Horticultural Society* 103, 55–59.

Whiteside, J.O. (1981) Diagnosis of greasy spot based on experiences with this disease in Florida. *Proceedings of the International Society of Citriculture* 1, 336–340.

Whiteside, J.O. (1984) Infection of sweet orange fruit in Florida by a common biotype of *Elsinoe fawcettii*. *Proceedings of the International Society of Citriculture* 1, 343–346.

Whiteside, J.O., Garnsey, S.M. and Timmer, L.W. (1988) *Compendium of Citrus Diseases*. American Phytopathology Society, Minnesota, 80 pp.

Wilcox, D.A. and Davies, F.S. (1981) Temperature-dependent and diurnal root conductivities in two citrus rootstocks. *HortScience* 16, 303–305.

Williams, J.G.K., Kubelik, A.R., Livak, K.J., Rafalski, J.A. and Tingey, S.V. (1990) DNA polymorphisms amplified by arbitrary primers are useful as genetic markers. *Nucleic Acids Research* 18, 6531–6535.

Williamson, J.G. and Castle, W.S. (1989) A survey of Florida citrus nurseries. *Proceedings of the Florida State Horticultural Society* 102, 78–82.

Williamson, J.G., Castle, W.S. and Koch, K.E. (1992) Growth and ^{14}C-photosynthate allocation in citrus nursery trees subjected to one of three bud forcing methods. *Journal of the American Society for Horticultural Science* 117, 37–40.

Willis, L.E., Davies, F.S. and Graetz, D.A. (1990) Fertilization, nitrogen leaching and growth of young 'Hamlin' orange trees on two rootstocks. *Proceedings of the Florida State Horticultural Society* 103, 30–37.

Willis, L.E., Davies, F.S. and Graetz, D.A. (1991) Fertigation and growth of young 'Hamlin' orange trees in Florida. *HortScience* 26, 106–109.

Wilson, W.C. (1983) The use of exogenous plant growth regulators on *Citrus*. In: Nickell, L.G. (ed.), *Plant Growth Regulating Chemicals*. CRC Press, Florida, pp. 207–232.

Wondimagegnehu, M. and Singh, M. (1989) Benefits and problems of chemical weed control in citrus. *Review of Weed Science* 4, 59–70.

Wood, B.W., Tedders, W.L. and Reilly, C.C. (1988) Sooty mold fungus on pecan foliage suppresses light penetration and net photosynthesis. *HortScience* 23, 851–853

Wutscher, H.K. (1977) The influence of rootstocks on yield and quality of red grapefruit in Texas. *Proceedings of the International Society of Citriculture* 2, 526–529.

Wutscher, H.K. (1979) Citrus rootstocks. In: Janick, J. (ed.), *Horticultural Reviews*. AVI Publishing Co., Westport, Connecticut, pp. 230–269.

Wutscher, H.K. (1989) Alteration of fruit tree nutrition through rootstocks. *HortScience* 24, 578–584.

Wutscher, H.K. and Dube, D. (1977) Performance of young grapefruit on 20 rootstocks. *Journal of the American Society for Horticultural Sciences* 102, 267–270.

Wutscher, H.K. and Obreza, T.A. (1987) The effect of withholding Fe, Zn and Mn sprays on leaf nutrient levels, growth rate and yield of young 'Pineapple' orange trees. *Proceedings of the Florida State Horticultural Society* 100, 71–74.

Wutscher, H.K. and Shull, A.V. (1975) Yield, fruit quality, growth and leaf nutrient levels of 14-year-old grapefruit, *Citrus paradisi* Macf., tree on 21 rootstocks. *Journal of the American Society for Horticultural Science* 100, 290–294.

Xiang, C. and Roose, M.L. (1988) Frequency and characteristics of nucellar and zygotic seedlings in 12 citrus rootstocks. *Scientia Horticulturae* 37, 47–59.

Yager, E. (1977) Drip irrigation in citrus orchards. *Proceedings of the International Society of Citriculture* 1, 110–117.

Yelenosky, G. (1985) Cold hardiness in citrus. In: Janick, J. (ed.), *Horticultural Reviews*, Vol. VII. AVI Publishing Co., Westport, Connecticut, pp. 201–238.

Yelenosky, G. and Hearn, C.J. (1967) Cold damage to young mandarin-hybrid trees on different rootstocks in flatwoods soil. *Proceedings of the Florida State Horticultural Society* 80, 53–56.

Yelenosky, G. and Young, R. (1977) Cold hardiness of orange and grapefruit trees on different rootstocks during the 1977 freeze. *Proceedings of the Florida State Horticultural Society* 90, 49–53.

Young, R. (1969) Effect of freezing on the photosynthetic system in citrus. In: Chapman, H.D. (ed.), *Proceedings of the First International Citrus Symposium*. University of California Press, Berkeley, pp. 553–558.

Young, R.H. (1977) The effect of rootstocks on citrus cold hardiness. *Proceedings of the International Society of Citriculture* 2, 518–522.

Young, R.H., Albrigo, L.G., Cohen, M. and Castle, W.S. (1982) Rates of blight incidence in trees on Carrizo citrange and other rootstocks. *Proceedings of the Florida State Horticultural Society* 95, 76–78.

Zaragoza, S. and Alfonso, E. (1981) Citrus pruning in Spain. *Proceedings of the International Society of Citriculture* 1, 172–175.

INDEX